I0748455

JAINTIA ORAL NARRATIVES

The Author

Dr. S.N. Lamare was born in Chutwakhu, Jowai, Meghalaya and spent his early childhood days and schooling in Jowai and then went to Shillong for higher studies. He graduated from St. Edmunds College and completed his Master's and Doctorate from the North-Eastern Hill University. Dr. Lamare has been actively engaged in research and has a varied field of interest. His principal area of interest in his research is to understand the various aspects of society, culture and economy of the Jaintia people. Among his well known published books are ***Resistance Movements in North East India: The Jaintias of Meghalaya 1860-1863; Reflections: A Collection of Poems; The Jaintias: Studies in Society and Change*** **and** ***Crying Skies a Collection of Poems***.

Dr. Lamare has also published a number of articles and papers both in vernacular and English in research journals, edited books and also in the form of popular writings. The varied interests of Dr. Lamare are reflected in the use of his training in history in producing historical documentaries.

At present Dr. Shobhan N. Lamare is an Associate Professor in the Department of History, North Eastern Hill University, Shillong, Meghalaya.

JAINTIA ORAL NARRATIVES

S.N. Lamare

Associate Professor
Department of History
North Eastern Hill University
Shillong, Meghalaya

2016

Regency Publications

A Division of

Astral International Pvt. Ltd.

New Delhi – 110 002

Publisher's Note:

Every possible effort has been made to ensure that the information contained in this book is accurate at the time of going to press, and the publisher and author cannot accept responsibility for any errors or omissions, however caused. No responsibility for loss or damage occasioned to any person acting, or refraining from action, as a result of the material in this publication can be accepted by the editor, the publisher or the author. The Publisher is not associated with any product or vendor mentioned in the book. The contents of this work are intended to further general scientific research, understanding and discussion only. Readers should consult with a specialist where appropriate.

Every effort has been made to trace the owners of copyright material used in this book, if any. The author and the publisher will be grateful for any omission brought to their notice for acknowledgement in the future editions of the book.

Cataloging in Publication Data--DK
Courtesy: D.K. Agencies (P) Ltd. <docinfo@dkagencies.com>

Lamare, Shobhan N., 1969- **author.**
Jaintia oral narratives / S.N. Lamare.
pages cm

ISBN 978-93-5130-971-0 (International Edition)

1. Folklore--India--Jaintia Hills (District) 2. Oral tradition--India--Jaintia Hills (District) 3. Jaintia (Indic people)--Folklore. I. Title.

GR305.5.J35L36 2016 DDC 398.20954164 23

Published by : **Regency Publications**
A Division of
Astral International Pvt. Ltd.
– ISO 9001:2015 Certified Company –
4736/23, Ansari Road, Darya Ganj
New Delhi-110 002
Ph. 011-43549197, 23278134
E-mail: info@astralint.com
Website: www.astralint.com

Laser Typesetting : **Classic Computer Services,** Delhi - 110 035

Printed at : **Replika Press Pvt. Ltd.**

~ Dedication ~

To my parents

And

to

all story tellers

Preface

The thought of compiling the oral narratives of the people have proved to be important and demanding considering the fact that they have carved out a space in historical research. In today's context Oral Tradition and Oral History are regarded as an important source of information in building up the history of a group of people where nothing much is known about them prior to the colonial period because of the absence of writing.

The need to put down these information in a written form arises from the fact that many narratives have disappeared with the demise of the elders in the society. Furthermore, the traits of storytelling are now things of the past, more so with the invasion of the electronic media and the new digital age. The "Monologue" and "Dialogue" has made its exit very fast. The old narratives are now being replaced by Urban Legends which are more attractive to the new generation.

It is in the light of these new challenges that an attempt is being made to come out with the "Oral Narratives" not so much as to preserve the old in their fight against the new, but also to provide an opportunity for the reader to venture once again into the realm of the earlier period and also to understand the same with a rational mind. Some of the references do have historical connotations and this may provide a window for researchers to relook into those areas which are of historical importance.

It is also interesting to note how memory, age, gender, communication behaviours, and oral constructions had influenced peoples' behaviour at a given time. The question on human memory and its impact on the reliability and validity of the same will invite further enquiry on the psychology of memory and its application to these kinds of narratives. The effects of time, absent-mindedness, mental block, biasness, memory loss and others are some of the areas that one needs to bear in mind on the process of reconstructing the past events.

However with limitless imagination, these narratives do find an outlet in expressing the experiences and sharing it with the community and the people at large. It would be nice if these oral narratives carry beyond the printed text and into different art forms like that of performing art, reverberation, documentaries and others.

It will not be out of context to say here that Oral tradition and narratives provides important insights into the culture, religion, society, economy, politics, matrimonial relationships, art and may even raise important questions and interpretations. The new methodological approach towards research will provide an opportunity to the scholars to answer some of the basic questions raised in this text and this may prove challenging.

My acknowledgement goes to Mrs. Betty Laloo, Mr. Edmond Lamare and my friends and relatives who have provided valuable information while writing this book.

A special thanks goes to Mr. Benjamin Syiem, an accomplished artist, for working on the cover of the book.

My thanks also go to Mr. Prateek Mittal, Astral International (P) Ltd., New Delhi for taking interest in the publication of this book.

I hope that the book will create an interest in the minds of the young researchers and it will add another dimension in the understanding of the Jaintia society.

On the whole I hope that the readers will find the narratives interesting.

S.N. Lamare

Contents

1
Introduction

In the case of some of the tribal societies of North East India oral history had been a part and parcel of the lives of the people and this was all the more important in the absence of writing. The importance of this kind of understanding can be seen from the fact that many of the stories about the origin, life and culture of the people can be seen in these kinds of narratives that are being shared from one generation to another by word of mouth.

Scholars and researchers of the 'modern oral history movements enjoy contemplating its origins, sometimes pointing out with excitement that all history was oral before the advent of writing'[1]. If one would try to study the life and culture of the ancient Greek, Persian, Chinese or for that matter that of the Africans, starting off with Herodotus, who gathered information of the Persian war in the fifth century BC, as well as Thucydides, who interrogated his witnesses to the Peloponnesian war[2] and many others, most of them have made use of the oral history and traditions in order to come to any kind of conclusion in their works. In the Zhou dynasty of China (112-256 BC), the emperor appointed scribes to record the saying of the people for the benefit of court historians[3]. Africanists points to the *groit* tradition in recording history, in which oral traditions have been handed down from generation to generation[4]. It was to this effect that Jan Vansina, a renowned historian and anthropologist pointed out that "ancient things remain in the ear"[5].

Oral tradition in a general sense would refer to the diffusion of cultural substance through verbal utterances. In this sense, especially in the case of pre-literate societies, these cultural substances would be diffused through talking or song and may take the form of folktales, myths and legends or any other. Despite the traditional prevalence of orally transmitted historical sources, such traditions fell into disfavour in the scientific movement of the late 19th century, and there arose a prejudice against oral history that remained strong for more than fifty years. The 19th

century German historian Leopold Von Ranke, protesting moralisation in history, said that the task of the historian was "simply to show how it really was"[6] and other historians enthusiastically took up the cause. Some historians, however, were never won over by the scientific approach.[7]

Having said the above we are now exposed to the question as to how do we define oral history? Many would try to understand it as spoken stories about things that happened in the past. Oral history is a generic term that may be interpreted in many ways. It would refer to a basic structured collection of spoken firsthand memories in an interview setting. Oral history is characterised by a structured, systematic planning process, thorough research, a careful emphasis on the depths and details of information collected, and adherence to strict processing techniques. Oral history, despite the generic use of the term, is a research methodology with a precise bounded meaning and a process that supports and defines the interview as the active collecting step[8]. In this sense it can be pointed out that recorded speeches or someone reading a historical document and recording it is not oral history. Furthermore, audio or video reminiscences collected by asking some older person to talk of the olden days cannot be regarded as oral history until and unless proper research work is carried out on the said subject.

Allen Nevins had introduced oral history as a name for the spoken memory interviews when he began his work at Columbia University in the late 1940s. Since that time the term has been identified with the process of collecting oral information about the past[9]. In the years since the term first began to be used, oral history has come to have a popular vernacular and an archival meaning[10].

THE CASE OF NORTH EAST AND THE JAINTIAS IN PARTICULAR

One important aspects of oral history is when the traditions of the people are being used in order to reconstruct the past. In this sense oral traditions can also be described as living stories. The collections and use of these oral traditions is still very strong especially in the case of the indigenous people of North East India. Oral traditions in most cases are being told in the first person, but then these stories can go to the distant past. Depending on time and space, these traditions can also include the sacred and the secular and would reflect much on the manner as to how the story is to be told or narrated.

In the case of the societies of North East India, these traditions and oral narratives has been providing a kind of insight into the past and in so many ways would make history come alive. It helps us to understand not just what happened but also offers an opportunity to uncover the layers of meanings embedded in those stories and also to interpret the past. Oral history can also help preserve languages and dialects. By preserving the sound and cadence of spoken words, it can help keep languages alive. It is often used as a tool to save the vernacular and also for documenting the lives of ordinary people.

The other aspect of it lies in the fact that many of the oral narratives, originated at a time when man had no valid answers to many questions of existence[11]. The natural phenomena which man experiences in his everyday existence, talks much about the kind of influences that nature was having on their thinking processes. The attempts to answer many of these perplex question would require the individual to create a story so as to confer powers on God or other deities and also to give place to semi-divine heroes within their own context. The manner in which these stories are being told and narrated especially during the time of festivals and rituals would give then some form of sanctity over the years and this with time becomes part and parcel of the life and culture of the people in the society.

Today there is a kind of general acceptance especially among the literate and illiterate sections of the different societies of North East India and the Jaintias of Meghalaya in particular, to understand the functions and importance of these traditions. Attempts are also being made in the larger context to place these narratives in a different kind of understanding and to look into the wisdoms of the people of the past there in. Just as is the case with the different tribal groups of the country, the indigenous people are very close to nature, and many of their social customs and practices or their way of life is being inspired by the things that are there around them. But then, as with any other oral traditions, here also one can see the elements of exaggerations working overtime and a lot of stress is being given on the farsightedness of the older generations without looking into the kind of limitations and world view in which the people of the past were placed in. There is no doubt that many of these stories would revolve around things and objects seen in nature and how the same would be woven into a story and narrating the human experience. One such area where it would amuse a reader is the kind of belief which is still very strong with the people where both man and beast were using the same language and were able to communicate and understand one other. This is strongly reflected in the story of *Ka Iaw Luri Lura* and how God later created confusion with their common language and in the process man, animals and birds came up with their own.

The other kind of assertion would speak of the kind of hidden meanings and advices that these stories are having. Today the neo-literate class would dwell on these themes so as to make a claim that these stories are not meant merely for amusements but also for the purpose of giving moral lessons, advice and instructions to the future generations, so much so that even the culture of the people is also being reflected through these narratives. There might be some element of truth here, but as mentioned earlier the locale and the mindset of the people should not be negated before coming to this kind of conclusion. The kind of revivalism that had taken place from the 1950's onwards, especially in the case of Meghalaya, reflect much on the kind of emotional outpourings and an attempt to understand these parables of the past in a new light and, therefore, these kinds of conclusions. In the present day context when there is a lot of talk of environmental degradation and global warming, these stories and oral narratives have once again made their appearance in a big way, linking the understanding of the people towards the environment, through the preservation of the forest in the form of the sacred groves, forest, river and streams and relating them up with the appeasement of the malevolent and benevolent spirits who are roaming the hills, forests and also dwelling in the rivers and streams. The general description would focus on the capacity of the older generation to understand the importance of the natural surrounding and how to protect the same by attaching the sacred and the profane to them and in the process preventing man from polluting the same. This being accepted, one would be curious enough to know as to how the agricultural practices that is being carried out in the hills in the form of jhum cultivation and the impact that it had on the ecology, which at present is the major cry of the environmentalist, can be related to. One cannot deny the fact that, the situation would have been worst if it had not been for the scientific intervention that had taken place with the dawn of India's independence. But the situation is still grave in the case of the North-East just like it is in the case of the other tribal areas in the rest of India.

The supplementary aspects of these oral traditions can be seen when attempts were made by the ruling authorities in the past to see that these stories do perpetuate so as to continue and survive. It serves many purposes. One such area can be seen and understood is when an attempt is being made to link the mythological narratives with the institution of Kingship or for that matter the clan. In such a situation the theological and social implications of these narratives gains significance.

Having said the above one cannot deny the fact that in the present day context, oral traditions are being regarded as an essential source in understanding the tradition and culture of a group of people or a community. The implication of the same can be seen in the recent years when historians have tried to use the oral tradition side by side with their primary and secondary sources in studying a particular group of people or a community. Their importance lies in the fact that they could generate a lot of information, especially with matters relating to the pre-literate societies, where much of the tradition of the people had been buried in time, with little or no written records at all.

Jaintia Hills District is located in the easternmost part of Meghalaya. It is bounded in the east and north by North Cachar Hills District formerly known as Mikir Hills District and Karbi Anglong District of Assam respectively. In the west the District is surrounded by the Khasi Hills District and in the South by Bangladesh[12]. It has a geographical area of 3819 square kilometers. It is the second largest district occupying 17.03 percent of the area of Meghalaya[13]. It comprises two administrative civil divisions, Khliehriat and Amlarem. Jowai is the only town, which is also the district headquarter[14]. According to the 2001 census, the population of the Jaintia Hills District is 2,99,108 with a population density of 78 person per square kilometer[15].

The Jaintias are one of the many tribal groups of North East India who do not have a script of their own and for that matter any written history. Whatever little is known about them, is being handed over by their ancestors orally right from time immemorial. Though the people have come and stayed in this beautiful landscape, the history and origins of the Jaintias or *Pnars* of the upland region (as this is the term they prefer to call themselves) is shrouded in mystery. In fact nothing is known about their ancestral lands, their migration, before they came over to these hills. Very little work had been carried out on them and the ones that have appeared in the form of articles and news items in journals and newspapers are being based on secondary sources only.

The Jaintias' history of the distant past is mixed up with myths, legends, folk narratives and these are being handed over from one generation to another. Stray references about the people can be traced back to the writings of the Ahom and Koch chroniclers and it goes back to the 15th and 16th century. It was only with the coming of the British that a more detail account about the people started to appear. This came about because of the fact that the officers posted in the hills had to write a report about the people as part of their administrative exercise. They on the other hand not only learned the different language of the area that they were posted in but at times even the local dialects. They at the same time started to document the old tradition of the people and this was made possible by laying stress on the oral tradition which the people had managed to preserve from one generation to another.

CONCEPTUAL FRAMEWORK

One question that is been generally asked is, whether these narratives can possess any elements of truths after all these years of being transmitted verbally. This is one area that needs to be looked into carefully. Any kind of oral transmission would depend much on the memory of the narrator, and since this would again involved the mind, there is every possibility that the facts and events may get distorted and, therefore, the conclusion be subjective. Oral historians have long been concerned with issues of memory, particularly how people remember and what shapes their memories. Memory is the capacity to store experience and then to recall or retrieve it. It is obviously essential to our ability to function[16].

There are three basic classifications of memory that are accepted by most cognitive (perceiving or conceiving) psychologists: short-term memory, sensory memory, and long term memory[17].

Short term memory is the system or systems that enable one to store and retrieve information for a short interval, *e.g.* Storing and retrieving a telephone number only for the length of time required to dial the number.

Sensory memory is dependent upon visual, auditory or tactile sensation, taste and smell. These perceptions may find expression either in short-term or long term memory.

In an informative book, *The Seven Sins of Memory*, published in 2001, Daniel L. Schacter provides a useful amalgam of the recent findings by describing the various ways that memory can become distorted or otherwise fail[18].

It is important to remember that memory is a complex personal experience. Studies of localised brain activity reveal that neither perception nor memory is a unitary thing. During visual perception, for example, the colour, shape, and motion of an object are processed in widely separated parts of the brain. Similarly, during the act of remembering, various widely separated parts of the brain light up (become active), depending on whether the memory is for sound, a conversation, or an image[19].

It should come as no surprise, therefore, that various kinds of brain injury produce quite specific effects on memory as well as on perception. Nor should it come as a surprise that there is an intimate relationship between perception and memory.

According to Daniel L. Schacter, the seven sins of memory are transience, absentmindedness, blocking, misattribution, suggestibility, bias and persistence[20].

Transience would refer to how memory changes with the passage of time. At relatively early points on the forgetting curve - minutes, hours and days, sometimes more-memory preserves a relatively detailed record, allowing us to reproduce the past with reasonable if not perfect accuracy. But with the passing of time, the particulars fade and opportunities multiply for interference-generated by later, similar experiences-to blur our recollection. We thus rely more on our memories of the gist of what happened, or what usually happens, and attempt to reconstruct the rest by inference and even sheer guesswork. Transience involves a gradual switch from reproductive and specific recollections to reconstructive and more general descriptions. Example, attending a party and trying to recall the people, the clothes, the colour, the content of conversation and how the forgetting was exhibited by memories that are blurred with the passage of time[21].

Absentmindedness on the other hand is common with people who are very busy. It is exemplified by the failure to readily locate one's keys. This happens because of the failure to recall, to encode and previously to store the memory of where the keys were discarded. Schacter points out that this kind of failure usually occurs because one is busy initiating some other task at a time when he or she should be encoding and storing the memory of the key's location[22].

Another common experience with age is blocking. This can be attributed as another vice. It is the inability to recall the name of a person whom one might know quite well. The Oral History Association's *Evaluation Guidelines* call attention to this problem by indicating that oral history interviewers should take care to use skills appropriate to the interviewee's condition like health, memory, mental alertness, ability to communicate, time schedule *etc.*

Misattribution is another problem which occurs when one fails to recall properly the source of recollection or memory. One common example of this problem is manifested when, having seen pictures of oneself as a child, one may come to think that he/she remembers the event being photograph. If pressed, however, the person can provide no accurate information about the event other than what is visible in the picture.

Experiments on the effect upon memory of providing false or misleading information have been carried out from time to time. It is well known that leading questions can colour memories and under some circumstances induce false memories. Suggestibility would mean when false information would be embedded in the questions. This false information has a powerful influence on the mind more so if these information was supplied by an authority figure *e.g.*, of cross examination by the cops or in the court. But then again there are people who are able to resist misleading external influence upon their memory. They can be treated as valuable participants in providing reliable information[23].

The sin of biasness refers to distorting influences of personal knowledge, beliefs, and feelings on new experiences or later memories of them. Just as what are likely to be seen is determined by the interests and biases, what is being remembered can be influenced by the same kind of interests and biases[24].

Persistence is another problem that one faces. There are some memories which refuse to fade, and when this occurs the memory tends to involve regret or trauma or some other negative emotion. All emotions tend to strengthen a memory, but negative emotions can strengthen the memory to a point where it is intrusive and interferes with normal activity. The amygdale (a brain center that becomes active during fear) and related structures contribute to the persistence of such experiences (fear) by modulating memory formation, sometimes resulting in memories we wish we could forget[25].

As can be seen in the recent past, many of the oral traditions of the people would be handed over to the next generation by the village elders or a story teller or at times even by the maternal uncle or aunt who has the knack of doing the same within the family. It is here where the authenticity of these traditions is being put to the test and there is no guarantee that they would not be modified or simplified with the needs of the time. As it can be seen today, there are a number of occasions when a simple event or a narrative would be given colour and be taken to great heights, in the process glorifying it. This kind of assertion may occur, because the situation may have demanded and necessitated such a development or just because to fulfill the needs and aspiration of few individuals with ulterior motives in order to gain something out of it. If this kind of development is allowed then the superlative coating that is being given over a particular event or tradition, over the years will take the people far away from the truth. This perhaps can be said to be one of the major danger and drawbacks in understanding the oral traditions of the pre-literate societies.

The problem of understanding the oral tradition is compounded further, when a researcher do not know or understand the language of the people in study. It has been noticed of late that a host of researchers would carry out research works without looking into one important criteria of research and that is to know the language of the people in study. This have been conveniently neglected or purposefully avoided by taking the help of the interpreter. The danger lies in the fact that the researcher will have to depend entirely on the information provided by the interpreter, who, interestingly again, more often than not, fails to provide the correct information given by the people. Worst still is the kind of generalisation that is being given by the interpreter and in the process many of the striking features of the people in study would go amiss. Furthermore the interpreter may think and pass his own judgement over the information provided and may or may not consider them to be irrelevant or at times considering the information to be incorrect and in the process correcting them. So, therefore, the conclusion that is being drawn from this kind of works can be anybody's guess. It is therefore, the duty of the local historians of the region to give special attention to this area of study by using new tools and methodology and to contribute something to this huge corpus of study. As for a historian, the oral traditions cannot stand independently as a source of information, nor can they be treated as truths. "It is not possible, therefore, to rely solely on oral testimony for historical research".[26] The oral traditions will have to be corroborated or even contradicted and crossed check with other sources, whenever possible before reaching on a conclusion.

The Jaintias have a very strong oral tradition, which is being handed over till today to the present generations. This tradition also talks about their origin and refers to their *U Khadynru Wasa* or the Sixteen Huts who stayed in heaven.[27] From this *Khadynru Wasa, U Tre Kirot* or God had ordered the *Niaw Wasa* or the Seven Huts to go down to a special place prepared for them *i.e.*, the Earth.[28] In the process, out of the sixteen, nine remained in heaven and seven came to live and prosper in this earth. This tradition like many other traditions claims that men was directly created and send by God, to this earth. It has no room whatsoever to the theory of evolution. The people still believe that the *Niaw Wasa* came leaving the rest in heaven. This tradition is heavily coloured with myths and the interpretations are vague. It does not throw a clear picture on the origin and migration of the people. These myths and beliefs play an important role in the life of the people. They throw a lot of light not only on their way of life, but also in their moral and spiritual well being and are connected to their fundamental ideas about life and the universe.

In the subsequent discussions some of the important oral tradition of the people is listed in the form of short narratives.

NOTES

1. Rebecca Sharpless, "The History of Oral History" in Thomas L. Charlton, Lois E. Myers, and Rebecca Sharpless (ed) *Thinking About Oral History: Theories and Applications*, p.19.

2. *Ibid.*
3. *Ibid.*
4. *Ibid.*
5. *Ibid.*
6. Rebecca Sharpless, op.cit., p.20.
7. *Ibid.*
8. Barbara W. Sommer and Mary Kay Quinlan, *The Oral History Manual*, p.2.
9. Barbara W. Sommer and Mary Kay Quinlan, op.cit., p. 1 .
10. Linda Shopes, "Background Paper: Oral History", in *Oral History Association* .
11. Soumen Sen, "Narrative, Ritual and Historical Events: The Jaintia Identity" in P.M. Passah and S. Sarma, *Jaintia Hill: Home of A Meghalaya Tribe-Its Environment, Land and People*, p. 88.
12. I.M. Simon, *Khasi and Jaintia Tales and Beliefs*, p. 1.
13. *Census of India 2001, Meghalaya-Jaintia Hills District*, Series-18, Part-A&B, Shillong, Meghalaya. .
14. *Ibid.* p. 5.
15. *Ibid.* p. xvii.
16. Alice M. Hoffman and Howard S. Hoffman, Memory Theory: Personal and Social in Thomas L. Charlton, Lois E. Myers, and Rebecca Sharpless (ed) *Thinking About Oral History: Theories and Applications*, p. 275.
17. *Ibid.*
18. Daniel L. Schacter, *The Seven Sins of Memory*, p.4.
19. Alice M. Hoffman and Howard S. Hoffman, op.cit, p.282.
20. *Ibid.*
21. *Ibid.*
22. Alice M. Hoffman and Howard S. Hoffman, op.cit, p.284.
23. Alice M. Hoffman and Howard S. Hoffman, op.cit, pp. 286-287.
24. *Ibid.*
25. Alice M. Hoffman and Howard S. Hoffman, op.cit, p.288.
26. R.K. Billorey, "Oral History In North East India" in *Proceedings of North East India History Association*, (henceforth N.E.I.H.A.) 2nd Session, 1981, p. 16.
27. Shobhan N. Lamare, *The Jaintias: Studies in Society and Change* p.3.
28. *Ibid.*

2
The Legend of *U Lakriah*

The *Niam-Tre* over the years have survived the onslaught of time and today can be said to be one of the vibrant and most happening religion in the hills of Meghalaya. Interestingly the study of the said religion cannot be obtained from any sacred text or scriptures, as the founders of this religion did not have anything in a written form, as is the case with many other religions of the world in the distant past, be it with the greater or the smaller tradition. The religion of the *Pnars* has been handed over from one generation to another by words of mouth and there is a clear threat here until and unless the members of this religious group do something so as to document and put the same in a written form. The danger lies in the fact that the oral tradition cannot survive in its pure form and there is always the threat of having one or more ideas assimilating and influencing the ideas of the narrator of such a tradition. This had happened elsewhere and the case of the *Pnars* is no exception. To assert that the oral tradition had survived all these years in its true form without any kind of dilution of its facts and figures is any ones guess. Today with the spread of modern methods of documentation it is already a high time for the members of this denomination to do something concrete. Though the *Seiñ Khynroo-Khyllood* of the *Seiñ Raij* had already made attempts to document, yet there are still a lot of other things to be done.

The legend of *U Lakriah* have to be looked and seen as an attempt to document one of the important tradition and beliefs of the Jaintias, which in so many ways have influenced the development of the religion of the people *i.e.*, the *Niam-Tre*. The oral tradition has this to offer:

U Tre-Kirot the creator of heaven and earth is the ultimate source of power. He had created the living with flesh and blood and who will have to go through the process of procreation and also who have to face death. This would apply to both man and beast. Life was given to all his creation. The earth was filled with vegetations, drinking water and land for his creation to produce and multiply. But before this, in heaven stayed the sixteen huts. *U Tre-Kirot* wanted that Seven out of that Sixteen should go down to earth and to live and multiply in the place specially created for them. In order to finish the task that he had in mind, *U Tre-Kirot* chose one of the members from the Seven huts and spoke to him about his designs. He was *U Lakriah. U Tre-Kirot* appeared before him accompanied by thunder and lightning and in the form of a rainbow spoke to *U Lakriah.* The man was taken by surprise and fear, and listened carefully to the commandment that was put forward by God himself. He told him apart from the many that he should go to his own people and to tell them to get ready to go to a place that is being prepared for them by *U Tre-Kirot* himself. After the meeting with *U Tre-Kirot, U Lakriah* started to visit his people and told them about his chance encounter with God. There were many who doubted him and there were again many who believed in what he had to say.

Time went by and *U Lakriah* carried on with his usual work. One day while at work a sudden burst of thunder and lightning startled him and *U Tre-Kirot* appeared before him once again and asked him about the response of the people. *U Lakriah* who was again overtaken by fear told him that there were many who had listen to him and a few were still doubtful about the entire thing. He was then informed to go back to those who were not sure and to convince them and to get ready for the time for them to leave is almost near. With this few words *U Tre-Kirot* disappeared. A number of questions were raised by the Seven Huts on *U Lakriah* as to the day of their departure and whether there would be any sign or indication as to the time when they should leave the place. *U Lakriah* told them that he does not have any answer to all their queries and instead told them to be ready as being told by *U Tre-Kirot.*

On one morning *U Tre-Kirot* appeared before *U Lakriah* for the third time in all his glory and told him to inform his people to wait for the rainbow who would be guiding them. Soon enough all the people had gathered near *U Lakriah's* house and the rainbow started to move in a particular direction and the huge crowd followed from behind. The route was difficult and they passed through different "strata" until they have reached a particular place where there was a huge tree that was connecting to the earth or *Pyrthai.* This tree as said by *U Lakriah* was *Ka- Tangnoob-Ka-Tangjri.* When they reach earth the rainbow refused to move further as an indication for the people to remain and stay in that place. After this the rainbow disappeared suddenly.

The people started to build their houses but to their surprise anything that was constructed by them would fall apart by morning, because of the strong wind that was blowing over that place much to the agony of the people both young and old. Time went by and there was no indication from *U Tre-Kirot* as to what they should do next. This created a lot of inconveniences to the people and many of them were angry with *U Lakriah* questioning his statement about the promised land where they would live peacefully and ruled by themselves, without any kind of hardship. *U Lakriah* as before did not have any answers to all the questions put forward and instead told his people to be patient and wait for the sign from *U Tre-Kirot*. But time was not permitting the people to remain that way for long, as because this time they were made to face another major problem and that was hunger. A meeting was called and it was decided by *U Lakriah* till the time when *U Tre-Kirot* would appear to them, the members of the Seven Huts would be going to heaven during the day to work in the houses of the rich and to come down by night so that they can earned and not to die out of hunger. He also warned them that they couldn't stay overnight in heaven.

Initially there was a sense of joy among the people that they can go back to work and they can survive, but this was not to last long as because the task of going up and down every day was not easy. Many of them started to complain once again and some even challenge the authority of *U Lakriah* to prevent them from staying in heaven and to work there. A sizeable number of them went against the order of *U Lakriah* and worked and stayed in heaven. This had brought about a sense of distrust among the Seven huts and they were no longer united. *U Lakriah* pleaded for *U Tre-Kirot* to come and appear before the people and to explain to them, but to no avail. He then decided to fast and bathe himself with ash until *U Tre-Kirot* appears before him. Days passed by but nothing happened. This lasted for eight days and on the ninth day, in the early hours of the morning, there was a sudden burst and *U Tre-Kirot* appeared before *U Lakriah* who by this time was already covered with ash in a sitting posture. *U Lakriah* told him about his displeasure about whatever that had happened. *U Tre-Kirot* was listening to him patiently and finally told him that, "if he would give everything that is there with him to man he would still be unsatisfied and would crave for more". He told him further that before he send them to earth he had already pre-determined the things for man but he did not pre-determined his wants and therefore if he would be given all that is there in heaven he would still be unhappy. He also informed *Lakriah* that man would always remain unsatisfied with his wants and desires. After saying this he ordered *Lakriah* to go and call all his people so that he can talk to them directly. This was for the first time that God talked to the people directly and not through his messenger, *U Lakriah*.

A durbar was held and *U Tre-Kirot* talked to the masses with a commandment that runs as follows:

That the earth is being created with love and that his (*Tre-Kirot*) love is endless. This love is being personified in the pair that is chosen so that the future generation could carry on.

All souls belong to him and that man has to pass through the process of birth and death.

The universe is being made in three strata. The top most is called *Ka Soorkep*, where *U Tre-Kirot* and his fairies or *Ki Puri Blai* are staying. It is a place where the souls would rest for eternity and birth and death does not exist.

The second layer is called *I Bnein?*, where the Sixteen Huts were residing.

The third stratum is called *Ka Sla Khyndaw Pyrthai,* where the seven huts were taken and led by *U Lakriah,* under the order of *U Tre-Kirot.* After this only Nine would be staying in the second layer or *I Bneiñ.*

The Seven Huts were also to remember that throughout their generation, lineage should be taken from their mother and their clan name should remain.

The Seven Huts were to live a life of righteousness.

After their journey in earth, they would return to *U Tre-Kirot* through fire, and they were being warned not to bury anything on *Ka Rym-Aw* or mother earth by something that would decay and putrefy, so as to pollute it.

It was also commanded that whatever that is in heaven cannot take place or chance on Earth, and at the same time anything on earth 'let it be something new'.

This was the *Hukum* or the order of *U Tre-Kirot,* and after all these the earth became a beautiful place for the seven huts to stay, which they did and multiplied till the present day.

3

The Clan Cluster: *Ka Kur Soo Kpoh-Khad-Ar-Warnai*

Going by the oral tradition that has been handed over from one generation to another, a references is being drawn to the story of a man by the name of U Niang Langdoh, who along with his sister Ka long Langdoh landed in Jowai after days of travelling when they were compelled to leave their ancestral place because of the plague that was going on in their area. No information is being provided from these traditions about the place from where they had come. It was said that they were the first to come and stay in Jowai. After staying and being there for some time, he found out that the valleys were quiet fertile and carried out cultivation. As time went by he started to visit the hats of the neighbouring regions and came across a number of people and invited them to come to Jowai. He tried to lure them by promising good lands and high post to those who would come first. With time the number of people increased and U Niang Langdoh called this new place Jwai.

As days went by, the people were showing sign of prosperity and well being. As a sign of thanks giving and also to prevent the place and people from any kind of unforeseen calamity, U Niang Langdoh proposed that a ritual should be carried out. He called this ceremony *Ka Niam Behdieñ Khlam*, so that the entire place would be free from any kind of plague or diseases. This ritual was to be carried out once a year and he was supported in this attempt by all the others who had come to stay in Jwai.

U Niang and his sister Ka Long along with her four daughters Ka Bon, Ka Tein, Ka Wet and Ka Doh stayed in a place called *U Loom Soo Iung*. (To this there is another tradition, which says that the mother of the four sisters was *Ka Rangkit* and is being regarded till today as *Ka Sein Jait* or the mother progenitor or *Ka Bei* of the *Soo-Kpoh*). Going by the first version of the story, the *Loom Soo Iung* is a place near the present day K.J.P. Hospital. The oral tradition is silent about the husband of Ka Long Langdoh or whether U Niang ever got married or not. But the tradition has this to offer that as the place grew and the number of people increased, the four daughters of Ka Long also went to different areas to stay. Ka Bon went and stayed in a place called Chilliang Raij or the area of the children of the Raij. Ka Tein? went and stay in Lulong, Ka Wet stayed in a place called Wah Synji, presently called Panaliar and Ka Doh in Loom Iongkjam. It was said that during the *Behdieñkhlam* festival the four sisters would play a very important role and they would carry out all the ritual and practices.

In talking about the *Kur Soo-Kpoh-Khad-ar-Warnai*, they came into existence when the four sisters got married to people from different clans over a period of time. These four they came to be known as *Ki Iaw Bei* (or the first mother of a particular clan). Ka Bon got married to a man from the *Thangkhiew* clan and from this union we have the clans *Pasubon, Lipon* and *Rangad.* Ka Tein got married to a man from the *Mukhim* clan and from them the clan *Pakyntein, Nikhla,* and *War* came about. Ka Wet got married to a man from the *Shadap Manar* clan and out of this the clans *Paswet, Lakiang, Mutyen, Kma, Kynjin, Katphoh, Seinphoh, Leinphoh, Niangphoh, Lanong, Liwait* and *Litan* came into existence. As we can see from Ka Wet, we have twelve clans, which have their own *Moo-Tyllein?* or bone repository, in *Ka Blai,* below the hillock *U Lum Moo-Likso,* in Iongpiah, Jowai. It is again because of these twelve clans that it came to be known as *Ki Khad-ar Kur,* and presently known as *Ki Khad-ar warnai.* Ka Doh on the other hand got married to a man from the *Dkhar* clan and out of which the *Langdoh, Siangbood, Syngkon* and *Nangbah* came about. All the above belonged to the clan cluster *Soo-Kpoh-Khad-Ar-Warnai* and no marriage can take place within these clans. Reference is also being drawn to the clans, *Rymbai, Najiar* and *Toi,* where collectively they belong to the clan cluster *Ki Le Kyllung.* It was said that in the past a lady who was known as *Ka Piah Rah* who started the clan cluster of *Ki Le Kyllung,* had to breast feed a baby whose mother died at child birth, who happened to be from the Langdoh clan, *i.e.* tracing their lineage to *Ka Doh.* This act on her part made the members of the Langdoh clan to express their gratitude by coming closer to them and it was ordained that the members of the *Le Kyllung* cannot enter into any matrimony with the clan members from the linage of *Ka Doh.* On the other hand they were free to inter-marry with the clans that came out from Ka Bon, Ka Teiñ and Ka Wet. It was because of this clan relationship that the members of the *Soo-Kpoh-Khad-Ar-Warnai* are being considered to be the original settlers and they have the right over the administration of the place.

Apart from the above, references are also being drawn to the existence of the *San Syngkhong* and the *Iaw Chibidi.* In the case of the *San Syngkhong,* we have the clans *Shylla, Pde, Pariat, Blah* and *Slong* coming together and tracing them from the same ancestress. The *Iaw Chibidi,* on the other hand is being regarded as the progenitor of the seven different clans like the *Laloo, Nartiang, Marong, Pyrbot, Kjam, Hek, Katkeh* and belongs to the same clan cluster.

Going by the tradition of the people, it is only the members of the *Soo-Kpoh-Khad-Ar-Warnai* who can take up important and responsible post in the field of administration in the Jowai *Elaka* or with other things relating to the religious ceremony of the *Niam-Tre*. In this sense it can be added that the first *Daloi* of Jowai was *U Dong Sniriang Paswet*, from the lineage of *Ka Wet*. The first *Pator* was *U Ksan Pasubon*, from the linage of *Ka Bon*. The first *Sangot* was *U Tamon Pakyntein* from the lineage of *Ka Teiñ* and the first *Langdoh* was *U Het Langdoh* from the lineage of *Ka Doh*.

References

U Hamsilet Syngkon, "Ka Kur Soo Kpoh-Khad-ar-Wyrnai" in *Golden Jubilee Souvenir, 1941-1991*, Govt. Boys' High School Jowai, 1991, pp. 69-70.

S.R. Pasweth, *Ka Kur Sookpoh Khad-Ar Wyrnai Wa I Bynta Ha Ka Chnong, Jwai*. Mr. Pasweth is a retired Inspector of Schools in Jowai, Jaintia Hills District, Meghalaya.

Interview with Dr. Omarlin Kyndiah on 21:10: 2003.

Information derived from Mr. Sanwatki Challam Research scholar in the Department of History, NEHU, Shillong.

4

Myth of the Divine Descent of the Jaintia Kings

The oral tradition about the river fairy *Ka Li Dakha* and *U Woh Ryndi* is still very strong with the *Pnars* of Jaintia hills. The Jaintias used to refer to their land as *Ka Ri Khadar Dalloi* or the land of Twelve *Dallois* or chiefs. The oral narration suggests that at the beginning everything was fine with the people and the chiefs. But as time passed there was a lot of enmity between the different chiefs so much so that it brought along with it a series of problems and ultimately into clashes with each other. There was a kind of intrinsic warfare between the people themselves and no one was really safe. This had brought along with it a lot of hardship to the common man and their movements were severely restricted because of the jealousy and enmity that had crept into the society.

Version 1

It was said that the goddess of water Kupli, who was again regarded as the supreme goddess from all the water deities was very much displeased with the turn of events and the rivalry that had come about with the different dallois. She prevailed upon them that she would be sending her son to govern and rule over them and that the dallois should also swear allegiance to him and that all kind of rivalry should come to an end. But the question as to how and when she would be sending her son was not specified by her.

The oral traditions narrates that at this time in a particular village called Sutnga there lived two sisters belonging to the Pala clan and who have reached menopause. It so happened that one of the sisters got pregnant even after crossing the age of conceiving and gave birth to a boy and called him U Loh Ryndi.[1] Since these two sisters they belong to a very small clan they decided to leave their ancestral place and went to the southern part of Jaintiahills and stayed in a place called San-Chnong near present Jarain, Umlatkur. It was said that the mother of U Loh Ryndi died in this place. U Loh Ryndi as he grew up started to carry out cultivation and this he did near the river Thloo-Moo-Wi. It was at this time that the goddess kupli decided to send her daughter Ka Li to this river. Tradition has it that the fish travelled from the river kupli to Waikhyrwi to Prang then to Myntdu and then finally reaching Thloo-Moo-Wi.

U Loh Ryndi apart from carrying out his cultivation use to spend a lot of his time fishing in the river by the side of his cultivation land. On one occasion, it was said that he spend the whole day trying to catch a fish but as luck would have it he failed to get even a single one. It was at the time when he was about to leave, to go home, that he caught a small fish and he was thinking that he would cook the same for his dinner the moment he reaches home. It so happened that this man was so tired that he forgot to cook the fish and instead had whatever that was there in the house and went to sleep. The next morning he got up as usual and went to his field to work forgetting all about the fish. Later in the evening when he came back home he was surprised to find out that his house was cleaned, utensils being washed and food being cooked for him. He was completely taken aback by these kinds of developments and immediately went out to enquire from his neighbours whether they were a party to any of these things that had happened in his house. But wherever he went the response was always negative. He the asked as to whether they have seen anyone coming to his house and doing the same and to this also the answer was in the negative. He then went back home and was perplexed by the entire developments. The next day he went to work again and on returning back he was surprised to find out that the same thing was repeated on the second day. This time he was determined to catch the person behind this entire thing. So the next morning he got his gears ready as usual and pretended to go to work just like any other day. He left the house and walks a distant and then hid himself. Towards noon he approached slowly to his house and peeped from the corner to see whether there was any kind of movements inside the house. It was at that time to his utter surprise he found out that a beautiful damsel started to emerge from the fish which he had kept in the corner, in the kitchen of his house which he had forgotten to cook. He just could not believe his eyes. The next thing that he saw was that the beautiful girl started to clean his house and prepare food for him. He came out of his hiding place and got hold of her and there was a tussle as the girl was taken by surprised and wanted to get back into her shell of scales. He prevented her by throwing the scales into the fire. He then asked her who she was and she responded by saying that she was Ka Li Dakha, the daughter of the goddess Kupli and was sent by her mother and that she had come a long way to meet him and to be his wife. He then accepted her as his wife and they lived happily.

With time Ka Li Dakha gave birth to a son and a daughter. The son was named U Iakor Sing, believed to be the original source of the river Kupli. The daughter was named as Ka Lasubon Ksiar. It was said that the lineage of the Jaintia ruler started off with them.[2]

Version II

U Loh Ryndi was a man who lives at Thloo-Moo-Wi. He like his other country men uses to go for fishing in order to pass their time. One day he caught a fish just like any other day, but on reaching home he had forgotten all about the fish. Days passed by and the thought of the catch had never crossed his mind. One evening on reaching home he was surprised to find out that all his daily chores had been attended too. He had an impression that it could be his relatives or his neighbours who had helped him with it and did not gave much thought about it. It was only when these things continued for days and when he could not get answers for the same from his relatives and friends that he was determined to catch the person who was instrumental for such acts.

The middle part of the narrative runs in the same line like the above, but has to add the following towards the end. This source says that U Loh Ryndi and ka Li Dakha decided to go to river and to call her relatives. But before doing so he went to inform his mother about the intention and requested her to look after the house. In the mean time while they were away his mother started to clean and broom the house. On their return Ka Li Dakha found out that the broom was placed carelessly in the veranda of the house and took it as a bad omen and refused to enter the house and instead went running back to the river and took a plunge. U Loh Ryndi was sad and dismayed for such acts and did not know what to do. One day U Loh Ryndi saw in his dream that the mermaid Ka Li Dakha had gone to Waikhyrwi in Sutnga. He went to that place to fish and accordingly caught a fish which later turned into a beautiful lady. U Loh Ryndi could recognised that she was Ka Li Dakha and he settled down with her and according to the version available, they had two daughters Ka Raputong and Ka Rapunga[3] and three sons U Shyngkhlein-am, U Bania-am and U Tetia Ksaw.[4] To this P.R.T. Gurdon adds that they had twelve daughters and a son[5]. The oral tradition also has it that when the children had grown up, U Loh Ryndi and ka Li Dakha decided to go back to the water and before doing that, U Loh Ryndi took his fishing rod and stuck it to the ground, out of which grew bamboos, the joints and leaves which grows upside down till the present day. Meanwhile in the village the elders and the people were discussing about the unnatural birth of the children from the water fairy and decided to recognise them as being blessed. They also decided to recognised U Shyngkhlein-am as their leader and after a long deliberation the different villages decided to come under one umbrella and form a nation with Sutnga playing an important role in the state formation. So going by this tradition, the first king of the Jaintias can be traced to the son of U Loh Ryndi and Ka Li Dakha.

Version III

According to another version, U Woh Ryndi, was a resident of Sutnga village, located in the south-eastern part of Jaintia hills, who was a bachelor, went for angling in a pool called "Ka Thwai Khyrwi". The day was not eventful for him and he managed to catch only a particular type of fish called *Ka dakha khlur* (dakha = fish; khlur = star); in other words a star fish (?) The legend goes on that on his reaching home he forgot all about the fish and carried out his daily activities as before. One day coming home from work, to his utter surprise, he found that his house was properly cleaned and his food was cooked for him. This went on for some days and U Woh Ryndi was determined to catch the person who was doing all these things in his absence. Pretending to go for his daily chores, he slipped out of the house and hid himself and was continuously monitoring if there was any kind of activity that was going on in his house. It was during that time that he saw the starfish changing shape and taking the form of a beautiful girl and doing all the household works. U Woh Ryndi caught hold of her and prevented her from returning into her original form. He later married the maiden and the Jaintia royal lineage also started with this.[6]

To this, there is another story, which runs and relates that U Woh Ryndi was a merchant who uses to travel to Jaintiapur to purchase or sell his wares. In one of those occasions he took along with him a Bengali maiden. He kept this as a secret and later he declared her as a maiden obtained from the stream and later gained the popular favour of the local people who accepted her as their queen.[7]

Version IV

A man from the Tariang clan whose name was Loh Ryndi went for fishing at the Amwi river and caught a fish. It so happened that he forgot all about the fish and went on with his usual task. One fine day when he came back from work he found out that his house was cleaned and was very much in order. He was very much surprised by the turn of events and decided to keep a close watch the next day as to who was responsible for this act of kindness in his absence.

To his astonishment he saw a beautiful girl coming out of the fish and started cleaning his house. He could not believe what he was seeing, but then he realized that this was no ordinary happening. He entered the house slowly and managed to take away the skin of the fish before the girl could enter into it and then burned it. The girl was so beautiful that she charmed him right away. He proposed to the girl that he wanted to marry her. The girl agreed to his proposal, but then put one condition and that she would go to her people and bring them to his house.

In the meantime, U Loh Ryndi informed his mother about the events and requested her to clean the house as many people would be coming from the girl side and that in the meantime he will have to go with her to meet her family and relatives. The mother did not take the request very well and instead in an angry mood told him that the girl is not a queen where she will have to do all these chores for her and her relatives. The girl on hearing this felt very hurt and fled away to Sutnga. Loh Ryndi on the other hand was searching for this girl form place to place and from stream to stream. He was very disturbed and would spend most of his time fishing in the hope that he would be able to catch the fish once again.

One fine day he was able to catch a fish and kept it in his usual place. In the morning he found out that his house was cleaned and swept. He knew it then he had caught the same fish and he called the girl Ka Li Dakha. Without wasting any more time, he married the girl and out of their union two daughters was born to them. The story goes on to say that very soon after this incident both the parents died and the two daughters had to fend for themselves.

The two girls remained in their ancestral place. It was said that the younger daughter was not keeping good health and that she was sickly. A man from a nearby village Nongbah came over and saw the plight of the little girl and asked permission from the elder sister if he could take and look after her. The elder sister did not hesitate on hearing this and hand over her little sister to the man from Nongbah village.

Days and months went by, but the little girl would not just get better. She remained to be sickly. On the third year, the man out of frustration decided to sell her off on a market day. (This is an important twist in the story as it talks about the selling of the girl in a market place. There had been a great deal of speculation as to whether slavery had ever existed with the Jaintias. Though from time to time in a form of a narrative there is always this strong reference to *Ki Broo* or the slaves. This is very much there in the stories of the people as to how in times of war the enemies would be captured and made to work. Whether this story would help or not in understanding this particular aspect of the Jaintia society is a question that anyone can ask and probe further). But then his intentions was thwarted when a tiger blocked his way as he was approaching Wahiajer another small village. He then had to return home with the girl without completing what he wanted to do. On another occasion, while going to Puriang he was crossed by a deer and this got hold of his attention and spends much of the day chasing the same.

This incident of animals crossing his paths twice was not taken seriously by him initially. But then with time he started to ponder if there could be something like a message or a reason that was being communicated to him, so as to prevent him from selling off the girl. The more he thought about it the more it disturbed him and then came to a conclusion that there must be something strange about the girl. He decided to take proper care of her and gave her a bath as a start. In doing so he found out that the girl had a flower mark on her hair with symbols of the Jaintia ruling family. He immediately realized that she was no ordinary girl and that she belonged to the royal family. From this time onwards there was a remarkable change in his approach towards the girl and he took special care about her health and with time she grew up to be a very fine lady.

Time rolled on and the day came when he had to inform her that she should get married. There were a number of suitors in the village itself, but the man decided to leave the same for the girl to decide and to choose with whom she would like to settle down. To this effect he made an announcement that he would give her marriage to whomsoever she might favour. Going by the tradition and practice in those days, a number of people came to the house and the girl was supposed to give a sign of her approval by giving *kwai* to the one whom she thinks would be the right person for her. This in a way seems to represent a kind of *swayambar* that was being followed by the royal family in the plains, minus the garland and that the same is being replaced by the *kwai* or the betel nut and leaf. This perhaps had a lot to do with the kind of ideological influences that was coming from the plains.

The girl took her time and avoiding the other suitors went straight to U Shitang and gave him the kwai. The declaration was made and her intention was clear and accordingly they were married. Since Shitang was from the Sutnga village, the girl followed him there. There is something different with this oral narration as the events in this narrative talk about a practice that which is contrary to the traditional practices of the people. Right from the ancient period till today, the practice is such that after marriage the man goes and lives with his wife in her house and not vice versa. But this story along with the reference to the *swayambar* would give a hinduised form of interpretation to it.

The oral narrative goes on to say that out of the wedlock a son was born to them and they named him U Markusain. The boy right from his childhood days was known for his boisterous behavior. Markusain was also accused to have killed a boy of Sutnga village and for which his father had to pay a very heavy price. In order to rectify his behavior, the boy was sent to Borkhat to stay with the chief U Ksinor Saitsner or the intestine washer. This chief was known to have strange magical powers and was very powerful in that manner. It was said that his powers lies in his intestine and from time to time he had to wash them in a special pond meant for the said purpose. This was done in secrecy and should any one spot him doing the same he would die on the spot. It was because of this feat that he had, that he was known to the people around him as the intestine washer. The boy, perhaps out of fear began to behave properly and very soon gained the good will of the chief and his wife. He also came to learn about the secret of the chief and was working out a plan as to how he could caught him in the act and then to capture his power.

In the meantime things went on normally and the boy was also growing up. An opportunity came to him one day when in the course of a conversion with the wife of the chief he came to know that U Kisnor was on his way to perform the ritual. He manage to steal himself into the place and when Kisnor was in the middle of his act, U Markusain came out of his hiding and stood in front of him after which U Kisnor died on the spot. Markusian immediately claimed authority over all the areas of the chief and then returned to Sutnga. This oral tradition talks about the authority of Markusian extending even to the plains. Historically it was during the time of Bor Kuhain that the sutnga dynasty and Jaintiapur came under one rule. Could it be then that Markusain and Bor Kuhain was one and the same person? It seems that there is a possibility when the name of an individual could get corrupted with the passage of time, especially during a time period when nothing is being put in a written form and that a lot was depended on words of mouth.

For our understanding of the tradition we will refer to Markusian as the ruler who have united both the hills and the plains under the same crown. The tradition has it that Markusian ruled for many years from Sutnga and was succeded by a number of rulers. One of the successors transferred the capital from Sutnga to Nartiang and had his palace constructed there. It was here in Nartiang that he met a good number of Brahmins from the plains and he befriended them.

The oral tradition has it that Markusain was very wise and witty. In his process of befriending the people of the plains, he also asked them whether he could get some territories in their area as could be encompassed by the skin of a cow. The people of the plains thinking that it would be a small area told him that it would not be a problem and promised to give him the land. Markusain on the specific day appointed went down to the plains to meet the people there and to get hold of the land as was promised to him by the people of the plains. It was said that Markusain had already cut the cows' skin into small pieces and started to spread the same over the length and breadth of the land. The people of the plains could not object to his act but remained silent spectators and in the process he managed to get a large chunk of land under his possession. This tradition does throw some light as to how Jaintiapur came under the possession of the Jaintia kings and how with time the capital was shifted from Nartiang to Jaintiapur. This oral tradition does not refer to any form of war or conquest that was made by the Jaintia kings and how the same was brought under their control, but instead it talks more about the kind of wit and diplomacy that was used by the Jaintia king in order to get possession of the plain territories.

Version V

The Jayantia Buranji, an Assamese chronicle, gives important and interesting accounts about the Jaintia rulers. According to this source, Jaintia was ruled by a line of Hindu kings, the last of whom were Kedareswar Roy, Dhaneswar Roy, Kandarpa Roy and Jayanta Roy. The legend offers that Jayanta Roy had no issue. He solicited the presiding goddess of the state to bless him with a son. In his dream the goddess visited him and promised him a daughter who was to be named Jayanti, and in due course a daughter was born to him. The Jaintia kingdom according to tradition derives its name from her. She was married to Landabhur, son of the royal priest. Jayanta Roy in his old age made over the reins of the government to his daughter Jayanti, who ascended the throne under the name Ranee Sing. Meanwhile, Landabur, the husband of Jayanti was expelled by the latter on account of some indecency committed by the former. Landabur took shelter under the roof of the Sutnga chief and later was adopted as a son. After the death of the chief, Landabur succeeded as the chief of Sutnga. Jayanti in the mean time became repentant about her conduct towards her husband and prayed to the goddess who appeared before her in a dream and promised that a female child would be born of her shadow. The child was to be thrown into the water and to be devoured by a fish and would ultimately become the wife of Landabur under the name of Mutchoduree. A child was duly born, thrown into the water and was devoured by a fish, which was caught by Landabur. The fish turned into a beautiful lady, which swept and cleaned the house in his absence. Landabur eventually managed to spy on the fish and caught the beautiful girl and married her. Out of their union, Bor Kuhain was born to them. Tradition has it that Landabur made war on Muhammad Sultan the governor of Sultanpur[8].

Notes

1 U Shaimon Pyrbot, "Ka Li Dakha Wa Ka Ri 12 Dolloi" in Souvenir Panaliar Jowai Yungwalieh pp 66,67, 1991.

2 *Ibid.*

3 Interview with Mr. N. Sutnga in Sutnga Village on 13:6:1994.

4 Homiwell Lyngdoh, *Ki Syiem Khasi Bad Synteng,* p. 12.

5 P.R.T. Gurdon, *The Khasis,* p.168.

6 S.N. Lamare, The Jaintias, Studies in Society and change, pp. 46,47.

7 *Ibid*; and also in U Primrose Gatphoh, *Ki Khanatang bad U Sier Lapalang*, p. 95.

8 S.N. Lamare, Op.cit pp. 45,46.

5

The Origin of the Place of Worship of Jayanti Devi

This tradition has a strong reference to an incident that had taken place in the plain areas of the Jaintia kingdom. The narrative refers to a group of boys who were playing by the side of a small stone pillar at a place called Phaljor. While playing one of the boys referred to a stone as a deity and tried to imitate the offering of sacrifices, a ritual which was observed by their elders. The children decided to distribute the work among them and some went to gather flowers while others were engaged in cleaning the place. After everything was done a priest was selected from the group and one small boy volunteered to be the sacrificial lamb. This was in form, an imitation to the kind of goat sacrifices that was carried out by the people in that area.

The children tried their best to imitate their elders as far as the rituals and other sacrifices are concerned. The boy who volunteered as a sacrificial lamb was taken to the place of sacrifice and a long blade of grass was used as a sword. To their horror the moment the boy was struck with the long blade of grass, his head was severed from his body. The mocking sacrifice had turned into a reality. The boy was in two pieces. Struck with fear and horror, the boys left the place as quickly as they could and went to inform their elders as to what had happened.

The incident created a lot of uneasiness with the people and the matter was reported to the king. On hearing about the incident the king visited the place and was accompanied by the head priest of the kingdom who was a renowned tantric believer. On reaching the place the priest performed a ritual and read out the omens. He declared that the stone that was used by the children as god was that of Jayanti Devi, the deity of the kingdom. On hearing this from the priest, the king wanted to transfer the stone to his capital but all efforts to move the stone met with failure as the stone could not be taken out or remove from its place. This turn of events gave a special place and sanctity to the stone all the more. The king then ordered a shrine to be built around the stone with an order that regular puja was to be carried out, with an annual sacrifice of human. It seems that with this the case of human sacrifice in the Jaintia kingdom became all the more pronounced.

6
Oral Notes on Jaintiapur

Version I

The oral tradition of the people has it that a man by the name of U Sing Kason, came all the way from Jaintiapur to Sutnga to meet the Jaintia King. During the course of the discussion he requested the king to attack Jaintiapur and to annex it to his kingdom. The king felt victim to the stories and the riches of the plain areas as narrated to him by U Sing. The king then called for a durbar as it was a customary practice to do before any important decision is to be made and it was here in the durbar that the resolution to march to the plains was taken. The king along with his followers moved out form Sutnga and proceeded towards Borkhat and halted for the night in this place.

In the morning, the king felt in love with what he saw. He was determined to acquire the vast extensive plains of Jaintiapur where wet rice cultivation was carried out. On reaching the low lying lands, the tradition offers that, the king met a lady who was the owner of the land and negotiated with her on the terms of acquiring the land. An agreement was reached where the king was to pay her in cowries and that the same should be in heaps, on the concave side of two shields. The plot of land on the other hand would be the size of the deer skin. The woman was taken aback by this offer and without any second thought she agreed and the amount was paid to her immediately. The king then cut the deer skin into small pieces and scattered them in different directions. The woman stood there as a mere spectator and could not do anything. By the time the deer skin was exhausted the king had already acquired a large territory for himself.

Version II

There is still another version on Jaintiapur but not very much in circulation these days. According to this story, a number of Jaintia hunters from Jowai had gone down hunting to the border areas and by night fall had managed to enter Jaintiapur to settle for the night. Since they could not get any game, they were restless and wanted to have some adventure. They thought of attacking the palace. They planned their strategy and accordingly carried out their plan. They were successful in their attempts and by morning the palace was under their control and the throne was seized. The message was then sent to the hills and much to the delight of the people and the ruling authority the plain areas of Jaintiapur was annexed to Sutnga.

Version III

This tradition would refer to the time of the first king of the Pnars by the name of Shahjer, whose name cannot be maintained chronologically. The king would often hear stories about the fertility of the soil and the riches of the plain areas from the people who would frequent his area from time to time. This had attracted his attention to a very great extent and he wanted to explore the possibilities of visiting the area of the plains along with his deputy, Mar Kuhain[1]. At this point it may be mentioned that the reference to Mar Kuhain and Mar Kusain as in the previous narrative above has reach an interesting part. It is still difficult to ascertain that whether they are the one and same person. The previous tradition would refer to the latter as being instrumental in uniting the hills and the plain portion of the Jaintia territory through his wit and wisdom. Whereas this narrative sees him as the deputy of the Sutnga king who went along with him, but again this story does not fail to acknowledge the kind of intelligence that Mar Kuhain was having. The story would reveal it all.

On the fixed day the king along with his deputy and a few of his followers left their kingdom and decided to go down the plains. They travelled through the Khatdum area and then towards Umtalen till they reached Borkhat[2]. It was here from Borkhat that the king so the vast expanse of the plain areas and thought that if that particular plain area could be annexed to his territory then the plain area can serve as the bread basket for his entire kingdom. With this thought in mind he proceeded towards Jaintiapur and decided to stay there for a few days and to gather information from the people without revealing their identity, so as not to attract any attention. This according to him was important and crucial in order to meet his objectives. During their stay they came to know that the entire area was under the control of a nawab. This perhaps would refer to the time of the Mughal rulers

The first task that was there in front of them was to get hold of an interpreter who would understand their language and who would also help them in striking a deal with the Nawab there. Very soon an interpreter was found and an appointment was sought with the Nawab and the intention was made known to him. The Nawab was happy to know that the King himself from the hills had come down to purchase land, but told him that for every piece of land the size of the shield, the king will have to pay one shield of currency. This was indeed a strange kind of agreement that the Nawab thought about. The king though surprised with this saying did not show it to the Nawab but instead asked for some time, so as to discuss the same with his people in his kingdom. Having said this they all retreat back to the hills and a durbar was called. Many were just aghast by the strange demands put forward by the Nawab. Many things were said in the meeting and the money for the same was collected in a huge shield. Now it was time to go to the plains and U Mar Kuhain was ready to meet and confront the Nawab with his sense of humour and intelligence.

The delegation from the hills who accompanied the king and U Mar Kuhain were greatly impressed by the kind of abundance and fertility which the plain land had to offer. U Mar Kuhain disappeared from the crowd and went in search for a cobbler who has the aptitude of cutting leather into fine strips. Till this time the majority of the people who accompanied them were unaware of the plans of U Mar Kuhain. In fact many of them were having reservation as to how the king could agree to purchase land in the plains according to the term set by the Nawab. It was not long before a able cobbler was found and the details of surveying the land was made. The Nawab was pleased to receive one big shield of currency and told the king and his men to spot the location where ever they would like to have. The Nawab decided to call for a celebration, and a grand feast was arranged. Strange thoughts were running in his mind as to what would the king from the hill do with such a small plot of land, that is the size of a shield, where it would not be fit even for a single man to sit or do anything, except standing. Meanwhile the preparation for the feast was going on in full swing and very soon everybody sat down to enjoy the same except Mar Kuhain who went missing.

Markuhain took the shield, which he had brought from the hills and which was made out of leather, and went straight to the cobbler and told him to cut thin strips of the same without disturbing the shape of the shield. The cobbler on the other hand did a fine job as was asked and was paid handsomely after his job was done. The celebration was over and now it was time to measure the land, as promised, the size of the shield. Mar Kuhain asked his men and that of the Nawab's to stand around the shield and to pull the same towards themselves. The Nawab at first did not understand as to what was going on, but then it was just a matter of time. He soon realized that as the men were pulling the sides of the shield, the same was expanding, without distorting the circumference, and therefore, very much within the agreement. The Nawab now was very much convinced that he was outwitted by the Pnars from the hills, but could not do anything, as the agreement was still standing strong and both the parties were determined to see that the same should be executed in its totality.

The oral tradition has to offer that by the time the men from both the sides stopped pulling, the Sutnga king had already embraced a huge area under him, all this was because of the efforts and wit of U Mar Kuhain. It was said that right from this time onwards, the king of Sutnga extended his territory to the plains and immediately issued instructions that a palace should be built so as to commemorate this event. Along with the palace, the king also started a market at Jaintiapur and many other things were introduced in the plain area as was prevalent in the hills, which was very much in line with the customs and traditions of the Pnars.

Interpretation

The versions mentioned above may have varied in the story line, but the content is the same. All the three tradition speaks about the manner as to how the plain areas came under the control of the Jaintias. It one would try to analyse the details of the versions, one may asks a number of questions on the validity of the stories. But then the details of the annexation of the plain territories by the Pnars from the hills had raised a number of queries. It is a historical fact again that the plain of Jaintiapur was under the control of the Jaintia ruler and that the king had his summer capital in Jaintiapur. But then the basic question remains as to how the plain tracts came under the influence of the Jaintias? Was there any war fought? Perhaps the account from the Jayantia Buranji, the Assamese chronical will help to throw some light, as mentioned in version V above.

The account about the Jaintia king's ancestry differed much in details without distorting the content. The stories at different stages confirm the connection of the ruler of the hills with that of the former rulers of the old Hindu Jayantia kingdom. From the above it would appear that Sutnga and Jaintia came under the same crown during the time of Bor Kuhain, the son of Landabur, when he had inherited Jaintiapur from Jayanti. He was known to have maintained a palace at Borkhat. A shrine of Lord Shiva near Syndai is also associated with him. He was succeeded by Parbat Roy. It was during the time of Bijoy Manik the fifth successor that the kingdom was consolidated and the authority of the Sutnga rulers in the plains was established. He followed the policy of war and subjugation and three important principalities like that of Charikatha, Faljur and Jaflong were brought under his control[3]. At a later stage he also integrated areas like that of Dhargram, Chatul, Kharil, Araikha, Chaura, Pachbagh, Baurbhag, Piyaingol, Barnafaud, Satbak, Mulagul, Bajeraj, Bardes and Paschimbhag[4] within his fold.

These myths and legends have served their own purposes especially when we relate them to the political relationship between the two ruling houses of that of Sutnga and Jayantia, where with time it united into one. From this context it is important to understand that, myths such as these had helped the Jaintia society to integrate not only the cognate clans and bordering tribes, but also the technologically more advance Bengalis who were inhabiting the plains[5]. This again had a significant impact, when the process saw the domination of the Brahmins in the affairs of the state. This had helped the king to consolidate his rule and culturally influenced all his subjects.

Notes

1. H. Elias, Ki Khanatang U Barim, p. 188.
2. *Ibid.*
3. S.N. Lamare, op.cit p.48.
4. J.B. Bhattacharjee, 'Brahmanical Myths, Royal Legitimation and the Jaintia State formation' in *The North Eastern Hill University, Journal of Social Sciences and Humanities*, Vol. III, No. 1, p. 47.
5. S.N. Lamare, op.cit p. 47.

7

The Attack on Jaintiapur and the Story of *U* Ren

There is an old narrative which suggests that during the time of King Bor Kuhain II (1731-1770), Jaintiapur came under the attack of the Muslims from the plains. This tradition also gives an insight into the ancestry of Bor Kuhain. It says that for quite some time, the women folk in the royal family failed to give birth to any offspring because of a curse that had fallen on them. After many years, a princess gave birth to a baby girl and it was during this time that a Muslim woman also gave birth to a male child. The elders of the royal family who were in desperate need of a male heir and successor to the throne adopted the baby boy and brought him up in the tradition of the royal family. He was known as Bor Kusain II and when he grew up, was given the reign of administration and ruled from 1731 to 1770. He had a long reign and in the last days he gave up his throne to his successor Chatra Sing and became an ascetic. It was also said that after his death, Borkusain was buried and not cremated, in contrast to the practice of the Jaintias, as he was a Muslim by birth.

The storming of Jaintiapur by the Muslims would revolve around the reign of Bor Kusain II. This invasion had caused a lot of panic with the people and more so with the king, who was finding it difficult to fight back. The Jaintiapur fort was surrounded by the Muslims and many of the warriors of the Jaintia king were killed in the process. It was at this point when everything seems lost, that the guards of the royal prison reported to the king about a prisoner by the name of U Ren, who wanted to go out and to fight the invaders. The king then gave the orders that he should be freed and to be taken to the place where the arms were being kept, mainly swords and spears, so as to allow him to make his own choice of weapons. U Ren was taken to the place but was disappointed with the arms. His search through the rubble helped him to discover a sword meant to be used only during the time of sacrifices for the deity of Jaintiapur and was called *Ka Wait Pakopati*. When he tried it out, he found out that it was good and that he would be able to use it effectively in his fight. He was let out and he started to attack the invaders singlehandedly and killed them left and right. Many tried to flee to Kuwain and Dacca. U Ren went after them and he got to kill some more on the way.

The killing spree of U Ren continued, and on his way back from Dacca, he started to challenge many of the territories of the king of Malngiang. There is some problem with the chronology of events here and this can be said to be one of the drawbacks of the oral traditions. In the previous writings on the case of the Malyngiang rulers of Madur Maskut, Homiwell Lyngdoh had put a rough estimate that the kingdom of Madur-Maskut was completely routed out by the Jaintias in the year 1651. This was the time when Josamanta Roy (1647-1660) was ruling over the Jaintia kingdom. The rulers and subjects of the Madur-Maskut kingdom had scattered throughout the Khasi and Jaintia territory right from that time. The present narrative provides an interesting version that Bor Kuhain II who ruled from 1731 to 1770, seems to have the kings of Malngiang as his immediate neighbour. This is something which is very contradictory. The story goes on that, U Ren attacked the territories of the Malangiang king like that of Sohbar, Nongjri, Tmar, Umniuh, Tyrngai, Mawpat, Nongryngkoh, Nohron, Iewlah, Iewlad (Dienglieng), Mawjem, Mawkalai, Lad Umiong, Bhoi Lymbong till Dymrea (Dimarua, Nowgong)[36] and nobody dared to challenge him. Perhaps he was one of those Mars or giants that were there in the Jaintia Hills, where very little is known about his exploits. U Ren then took his sword and rammed it into a banyan tree and it remained there. The exact location of this tree is not known and the sacrificial sword known as *Ka Pakopati* remained embedded in that banyan tree.

8
The Story of *U* Chitang

The narration of U Chitang relates to a place called *Ka Riat Iam Sier*, which is in-between Jowai and Sutnga. U Chitang was regarded as a wise and a good man at heart. The stories that revolve round him mentions that he had established a good number of villages and that he had won many a battles too. It is difficult to ascertain as to what extent this kind of assertion in the oral tradition would hold water. If the person had fought battles then would it mean that he was also serving under the Jaintia king? The oral tradition is silent about these details. Nevertheless, the oral tradition that is still in circulation in places like that of Wataw and the neighbouring areas relates that U Chitang had a confrontation with the Jaintia king Borkuhain[36] with regard to the opening up of markets and *dwar-luti* or roads in those areas. One tradition has it that the *Dwar* at Borkhat was opened after a long confrontation between the king and U Chitang. The king finally prevailed over U Chitang and opened the market and the road ways. The place is still important till today.

U Chitang was nicknamed by his people as *U Kup-Dhup* meaning a person who covers himself with the quilt. (*U*=he; *Kup*=cover; *Dhup*=quilt). It is said that his quilt was so heavy and would weight half a maund. The reason behind this was because he would carry his cowries and other riches that he would need in his daily transaction and business, and all these would be tug inside his quilt. Those who do not know him would pass snide remarks about his mannerism and behavior.

By profession U Chitang was a blacksmith and had his trade at a place called Tuber. He specialised himself in the making of agricultural implements like the hoe and dao and was supplying the same to the areas of Rymbai and Sutnga. It was because of his hard work and skills that he was able to achieve success and became rich in the course of time. He became an important person in the area in which he was residing and was respected too by some.

Tradition has it that U Chitang had two sisters whom he had to look after right from the time they were small. He loved them both but at the same time he also wanted to know as to who loved him the most. As they were growing up he was looking for an opportunity to test them and to know as to whom he could take into confidence. Interestingly, there is no indication whatsoever in the tradition about their parents and it is being assumed that they were orphans.

On one occasion he decided to go for hunting and invited his friends to join him. For the first few days they could not get anything but on one fine day he and his friends came across a deer in its prime and without wasting any time U Chitang loaded his gun and was able to bring it down. There was a celebration on this price catch. U Chitang decided to take this opportunity that he had been waiting for so long to test his two sisters and immediately sent a messenger to his house to inform them that he had shot a bull meant for religious rites, belonging to other people and that he was being captured. The messenger also informed, as being instructed by him, that until he pays for the damages he will be taken to the king's court and that will put him in deep trouble.

The messenger went first to the youngest sister and apprises her of the problems that U Chitang was in. She flatly refused to help the messenger in giving money and also added that she doesn't have the same and that she refuses to help him as he had committed such a crime. She further added that they should go to *Ka Kongmai* (elder sister) for she can help.

On the other hand when the elder sister came to know about the plight of her brother she cried and was desperate as to how she could help him and to save him. She took whatever money that she had and left immediately with the messenger. Before reaching the spot, from a distance she could see a large number of people had gathered, she started calling for her brother and requested aloud to the other male members not to harm her brother and to release him. Interestingly, on reaching the spot, his sister was taken aback on what she saw. Instead of the bull, as informed, she saw a huge deer lying dead and she just could not understand the meaning of the same. When she was told by her own brother about his intention she was taken by surprise. In the meantime the youngest sister came to know the details from the hunting party and was angry and disappointed. But U Chitang knew as to whom he could trust as he had also got the details from his messenger. It is being said that the name of the place *Ka Riat Iam Sier* came to be known and attached from this very incident.

On another occasion, U Chitang called both his sister and informed them that he had kept all his riches on the twelve wooden boxes that he had made and told them to feel free and to pick the same for themselves. The youngest sister took six boxes that were heavy and left the lighter ones for her elder sister. On reaching her house she opened her six boxes that she had taken and to her surprise she found out that they contained all the implements of a blacksmith and some stones locally known as *Moo-Khuri* or hearthstone that are being used for the trade. This made her very angry and upset and she took all that was given to her and threw them in a place which came to be known as *Ka Noh Khuri* or a place where the *Khuri* is being thrown. *Ka Kongmai* (elder sister) took the remaining six boxes and she got all the riches that were kept there. U Chitang added all that he possessed including gold and silver and left for good to a place called Rymbai along with his sister who stood by him.

One interesting development that had come about as a consequence of this incident was with the beliefs of the people. There is a strong tradition with the Jaintias on the importance and reverence shown to this *Moo-khuri* or hearthstone that is being used to felicitate cooking at home. In the olden days the people do belief that the goddess *Khuri* resides at the hearth and she would guard their houses and property. It was because of this kind of reverence that is being attached to these stones that the people are being warned from doing anything otherwise. If there should be anyone who would spit or show disrespect to these stones then the goddess would punish them with a particular disease relating to the intestine and indigestion. Treatment of this king of ailment would involve the local medicine man who will carry out a series of rituals and chants and more importantly reading the omens. In such cases a rooster would be sacrifice to appease the goddess.

Rituals towards the goddess *Khuri* is also carried out during the time of house warming and especially when a new hearth is built inside the house. These sacrifices are never carried outside the house but inside. The hearth is being cleaned from time to time and the belief is that the goddess *Khuri* resides there.

A common practice that can be seen is the offering of rice grains by the elders before their meals on these stones before the others could proceed in the partaking of the same.

With the changing times and the disappearance of the traditional hearth and the invasion of the new modes of cooking like electric heater and gas stoves, it seems that the importance and significance of this belief has also disappeared, much perhaps to the inconvenience of the gods and goddesses.

9

Malevolent Spells in the Jaintia Society

Apart from the many beliefs and practices of the people, one area that cannot escape the attention of the Jaintias is the presence of the "evil eyes" in their everyday life. The general belief is such that this kind of power is being possessed only by a limited few and out of fear; people would try to keep themselves away from those who possess such powers or even the members of such family. People having such powers would carry out offerings and rituals to satisfy the "spirits" that had been residing in their houses and at times would elevate the position of the "spirits" to that of a house goddess. This kind of practice is found to be common with other forms of tribal beliefs found elsewhere in different parts of the world and are being referred to as magic, witchcraft or even divination. Magic for that matter is believed to have been known and practiced by many societies and is said to have existed from as early as the Paleolithic period. To the people in study, they would generally use local terms like *ske, kymbat and taro* to these kinds of spells that is being used and that can be inflicted on others.

Magic can be malevolent or 'black' and well intended or 'white'. Accordingly, it can be inflicted to destroy the rival party or enemy or to be designed in such a way so as to cure an ailing person. In this sense these kind of belief is still very strong with the people till date and it is very difficult to get detail information as to how this form of practice came to be part and parcel of the social life of the people. Though there are no hard and fast rule, nor scientific methods to account or check the happenings, but once a while when such happenings do take place, the local people would resort to a local sorcerer for treatment as they belief that there is no cure for this kind of problem from pure and applied medicines that are available today.

The subsequent discussion will throw some light at least on the beliefs of the people about the presence of the 'evil eye' in the Jaintia society.

The Jaintias beliefs that a woman who possesses a malevolent eye can cause harm to them. The attributes that these women have are hereditary and as the power of the evil eye is from within, it cannot be learned. Hence, it is generally inherited or handed down from one generation to another, whether one likes it or not. This practice in these hills is invariably evil. This would mean that to the vast majority of people from outside it may pose as a witchcraft to which they are familiar with or have heard or read from books or other sources. These evil eyes can cause a problem by just being stared at, as in the case of *ske* or by just being remembered through things that they may have exchanged, borrowed or lend with other people like in the case of *Taro*. In the case of *kynbat* this kind of spell is generally being passed when eatables are being given by the person possessing the same to someone who was not even aware, at times even without the least intention of hurting or inflicting any damage. This again would mean that the person possessing these spells cannot control the same as it may go against their wishes too at times. In the case of *Ka ske* and *kynbat*, the person affected would fall sick and immediate attention have to be given before the matter gets worst. The local people have their own medicinal preparations with herbs and other things to combat this menace and they are proven to be more effective than the medicines found in the chemist shop.

In the case of *Ka Taro* once a person is affected, he or she would fall sick and would be in a state of delirium for a long time with no proper sense whatsoever.[7] At times it would seem as if it would be a case of insanity as the person would speak and behave in a very different manner. It was also notice that at times the person would speak a different tone all together, far different from his/her normal voice and may even narrate the incident that had led him/her being possessed by the evil eye. These kind of mental imbalance would lead the victim to do certain things which at normal times would have seemed next to impossible. At times a young girl would exhibit tremendous strength that it would require a good number of strong men to control her or to pin her down, but in many cases she would manage to break free from their clutches and would walk and run faster than them, which in ordinary situation would not be the case. There were occasions when the affected person would try to inflict harm on themselves, or destroy their clothes, walk naked without any concern for themselves or others.

It would take a couple of hours before the evil eye would leave the tired and almost half-dead body for some time, but to enter again and to do more harm. It may be pointed out that the evil spirit can enter and leave the possessed body at any time it wishes. This kind of encounter would leave the victim weak and pale and slowly leading to death. However the sickness can be cured if the woman possessing the malevolent eye touches the body of the sick person and orders the spirit to leave the body. If this is done, the person gets cured instantly.

Another interesting feature that is being said about this evil eye is that, the spirit would possess only the local people and could not have any effect on those from outside, i.e. the people of the plains or other people belonging to other tribes. From this it is clear that the activities of the malevolent spirits are localised and that local people and children are basically its victim. This on the other hand would mean that the adults are not always safe, as the malevolent eye can attack but the young and the old.

In order to combat this menace that had been there right from the past, the people of Jainta Hills have resorted to the art of discovering the powers of the unknown. This is mainly carried out through the process of divination. Here the inspirational form of divination is generally being carried out. The people believe that answers can be obtained through the intervention of a local sorcerer especially when fowls are sacrificed by him and the omens being read. Though there are reports that the patients do get some kind of relief when such acts of faith are being performed, but one cannot really say that whether it really helps in curing them from this kind of attacks by the evil spirit.

As it appears in the present daytime, there is no proper medicine to combat the effects of these evil eyes. If we go by the traditional narration, many of the families do not even know that such an evil deity resides with them. This would be made known only when someone is being attacked by the same and would speak 'in tongues' of the members of the family. At this stage it would be known publicly that particular families are possessing such deity. This in a way has created a problem to the other members of the family who, unknowingly, had inherited the same and wishes to do away with it. Local tradition has it that if one wishes to do away with the evil spirit, at the individual level; they can do by severing off ties with the parent family and at the same time surrendering all kinds of earthly possession belonging to the family. It further adds that on a particular day in the dead of night the person concern will have to leave the family house naked and then only one can severe ties with the spirit. As it stands now it seems that there are no proper methods to do away or to combat the evils of the same, and till such methods are derived, the people of Jaintia hills will still be victims of those eyes which prey at them from a distance!

Reference

1. Shobhan N. Lamare, 'The Evil Eye In Jaintia Belief' in *The Sentinel* 17:09:95.

10
Ka Syiem Latympang

The story of Ka Syiem (queen) Latympang had raised a lot of questions not only with the scholars and researchers, but even with the common people. A number of queries are being raised as to whether the events as suggested in the oral traditions of the people had really taken place or not. Going by the social order and way of life of the Jaintias, it is apparent that in the traditional Jaintia society there is no scope for women to take up the reigns of administration. The old saying stress much on the kind of calamity and misfortune that lies ahead if a woman would take part in politics and for that matter her duties is being confined to domestic chores only.

These kind of assertion with regard to women in the Jaintia society and the strong references about the Jaintia queen in their folk tales and narratives, do pose a strong challenge on the views and tradition of the people today. Would it mean that the society was quiet liberal in the distant past and became more conservative and bias, with favoritism towards the men folk, only later? Could it be that this kind of bias came about when the society was exposed to the ideas and philosophy of the plains? Or could it be that the story about Ka Latympang was just a tale and had no historical authenticity. These are some of the areas that have to be looked into carefully in order to understand the Pnar society properly. Today with the stress that is being given on oral traditions as a source material for the study of history of a particular group of people or a community, the local oral historians of the region would find it interesting to look into the matter in far greater detail and to throw more light on as to what had really happened. Till that time the story of ka Syiem Latympang runs as follows:

Ka Syiem Latympang was the ruler of a place called Manar in Jaintia hills. It was said that she was 'a great queen, a sagacious ruler and a great warrior'.[1] Quiet close to Manar is another old village Changpung and the silent understanding between the two villages is that there cannot be peace without war. Tradition holds it that the two parties were regularly involved in inter-tribal warfare and there was a lot of hatred between the two. Manar as a village was comparatively smaller from Changpung in terms of territory and military strength but the commitment and unity shown by the people made the task of the latter very difficult to overcome them. It was also said that the people of Manar had built a fort around their village and was guarded heavily.

Repeated attempts were made by the people of Changpung to storm the fort but under the able leadership of their queen the people of Manar were able to repulse any king of intrusion. Having failed many times in their attempts the people of Changpung worked out a plan and invited the people of Manar to settle their differences by signing a peace agreement on a particular time and place set by them. The people of Manar who were all ready tired of war and bloodshed welcomed the initiative taken by the people of Changpung and agreed to meet at a place called Thangskai which is in between the two villages. It was also agreed upon that both the sides should bring white goats as many as possible to offer as sacrifices to the gods as a mark to commemorate the occasion.

The day of appointment arrived and the people of Manar went to the proposed spot, but were surprised to see that non from the other side had turned up. They waited for a long time and it was only towards the afternoon that the people of Changpung came and many of their men were disguised in white and from a distance they represented the animals that they were supposed to be brought along. Meanwhile the people of Manar failed to realised the danger that was in store for them and by the time they did, it was already too and the people of changpung swarm over them and attacked them mercilessly. In the midst of the confusion one of the men from Manar managed to escape and went to inform their queen Ka Syiem Latympang. On hearing these developments she was taken aback and overcome with grief. She was compelled to leave Manar and took refuge on a mountain called U Lum Tiniang.[2] This mountain is located near a village called Moo-Ka-Iaw. The Oral tradition of the people does not throw any light whatsoever as to how many people had accompanied her to the mountain. To think that a queen would be going all by herself to a secluded place, is something again, rife with speculation. It is again difficult to imagine how a queen could leave her subjects at home, because of the death of a few men's folk, and that too at the mercy of the enemies. These are some of the questions that have to be ascertained.

Going by what has been handed over from the past it appears that the queen took refuge alone in the mountain and on reaching the top she took out her bow and arrow and after a ritual fired her arrow on a particular direction and concluding that where ever the arrow would get stuck she would consider that place as her new habitat. Having said that she let go off the string and the arrow landed on top of a huge flat stone. Legend has it that the stone cracked into four pieces. Today the people still called this place as *U Moo-Soo Tno* or a stone with four sides. This place is in Barato. Ka Latympang resided in this place. Tradition has it that she lived in this place all by herself and that she managed to carry her gold and other riches (?). References about her cultivation and her being always in doors are being drawn and are very contradictory. Is it possible for a single lady to carry out such heavy works at that time all by herself? Going by this description it would appear that some of her close associates must have accompanied her if not the entire populace. There could also be a possibility that the majority of her subjects wanted her to go into hiding with some as because she was their queen and they did not want that anything bad should fall on her. As for the majority they were ready to accept what was there in store for them. But this again is just a matter of speculation.

Time passed by and the news about a rich and beautiful lady started to circulate around that area but people were hesitant to approach her on those issues. Very soon Ka Latympang was able to befriend a young lady of poor countenance and both of them would spend a lot of time together but the secret of her being a queen was kept away from the other lady. The beauty and riches of Ka Latympang impressed the young lady very much.

Not very far from where she was staying was a cultivation field belonging to one U Mied Iarynsut.[3]

Oral tradition has it that the young man was not only tall and handsome to look at but was strong and hard working. Ka Latympang use to watch and admire his disciplined life and very soon fell in love with him. U Mied on the other hand was not aware of such intentions on the part of Ka Latympang, as he was deeply engrossed with his works day in and day out. Ka Latympang who could not hold on her feelings anymore, wanted to express her emotions but never got an opportunity. To this effect she worked out a plan without the knowledge of her friend and invited the latter to join her for angling in a nearby stream, close to where U Mied was working. After spending sometime in the stream with her friend, Ka Latympang started complaining that she was not feeling well and pretended to fall sick. It was at that time that she told her friend to go and call the man who was working in the nearby field to help and to carry her home as she was in terrible pain. The man on hearing this from the lady, without any hesitation rushed to the spot where Ka Latympang was lying and carried her home.

On reaching her house she refused to let go of his hand and expressed her feelings to him. She was able to convince him to stay with her to which he did. Legend has it that the bullocks that were tied to the plough by U Mied remained unattended for days altogether and in time they turned into stones. The story of the two stone bullocks which can still be seen till today, in this particular place, would relate to this incident. The story ends abruptly here and no references can be drawn as to what happened later.

Notes

1. Soumen sen, Queens In Khasi-Jaintia Tales: A Query in *Proceedings of North East India History Association, Eleventh Session, Imphal1990, p.61.*
2. U Primrose Gatphoh, *Ki Khanatang Bad U Sier Lapalang*, pp.7-12.
3. *Ibid.*

11

Rice Cultivation in Jaintia Hills: The Myth

The story about the origin of rice cultivation in the hills of North East India had added new dimension and understanding, especially with the works of the scholars and researchers from the discipline of Folklore and others. Interestingly the different communities of the North East would offer different kind of stories about the same and many would revolve round myths, which provide a lot of scope for research in this area.

One such story about the origin of rice cultivation comes from Jaintia Hills in the form of a myth. The oral tradition revolves around a boy by the name of *U Stiar Hali* (*U*=boy; *stiar*=raised earth to trap water in the field; *hali*=paddy field). The date, the location and the other settings of the story cannot be derived, but the tradition is still very much in circulation with the people especially in the rural areas. Reference is being drawn to one young lady who was in the family way and was due to give birth anytime. This young woman was in some errand with her mother and had to cross a woodland area, when she went into labour. The two women tried to figure out as to what would be the best way of tackling the situation. After a quick discussion, the young women pleaded to her mother to get help as she would not be in a position to continue with the journey nor to go back to their native place. The setting sun was a strong reminder for both of them that time was running out. The mother took the bold decision of leaving her daughter in the woods and proceeded to the village to get help and other essential things that are necessary as part of the preparation for the birth of the baby, like the *larnai* pots, clothes and others. It may be mentioned that the *larnai* pots were used by the people in the past, especially in the case of birth and death.

Meanwhile the condition of the young women in the jungle was not good. The labour pain had increased many folds and she was aware that she would not be able to hold on, for her mother or the others to be by her side during childbirth. With whatever little strength she could muster, she managed to drag herself into the thick grass which acted like cushion and to keep away from the village dusty tracks. The thought of delivering the baby all by herself and at the mercy of nature made her emotionally excited and at the same time scared. It was soon dark and it was in those moments when no one was around except for the sounds of the insects and a clear night sky above her that she gave birth to a baby boy.

It took her quiet sometime to recover from the stage that she was in and in the early hours of the morning she decided to make an attempt to proceed with her newborn baby towards the village. This intention was thwarted the moment she started to walk because of the extreme pain of childbirth. She sank to the ground, just like she stood up and came to realise that it was next to impossible to carry the child and to walk at the same time. The sense of survival got the better of her and with a heavy heart left the baby behind, in the jungle and went towards the village.

At this point the narrative takes a different turn altogether and shifts its focus on the newborn child. The narrative adds that after the child was left to its own fate, a tigress that was prowling in the area moved towards the direction of the baby. Curious enough to see a baby lying in that condition, instead of attacking, sniffed the baby and took it to its hiding place. The tigress took care of the baby along with her other cubs and the child grew in the ways and manners of the tigers.

Meanwhile the young woman reached her village and narrated the whole story to her mother and the family members and accordingly a search party left immediately to look for the newborn child. Though traces of what happened in the night were very clear, yet the baby was nowhere to be found. The baby seemed to have disappeared without a clue. After a few weeks the search was given up and all hope was lost to find the newborn baby alive.

The child who grew up mainly on the tigress milk proved to be strong and healthy and with time learned the ways of the tigers. The narrative adds that it was a different kind of world altogether where the boy was able to communicate with the animals around him. The tigress tried to convince the boy to return back to his people and to live the life of humans, but the boy who by this time had reached his teens argued back and referring to the difficult nature of human behaviour and lives, it would be very difficult for him to survive in that condition. He preferred the way of the tigers. The tigress then promised that she would find a way as to how the boy would sustain himself.

The oral expressions give information on the manner as to how the tigress taught the young man to cultivate paddy. A place was chosen and the boy was taught to prepare the paddy field and the seedbed to cultivate the same. The agricultural tools were provided to him by the tiger like the *dao* and the axe. When everything was done it was time for sowing, but the seeds were not available. The tiger managed to do the same by scaring a passer-by who was carrying a sack of rice grains and handed it over to the boy. The boy made use of the grains and started sowing it. The agricultural activity that was associated with sowing, weeding, and others compelled the boy to built a hut for himself near the paddy field and to start living all by himself and not with the tigers.

Regarding the rice grains or the seeds, it was narrated that at one point of time a man who was working in a swampy area for a different purpose altogether, after a few days saw a great number of saplings growing like wild weeds. He became curious about them as they did not appear to be like the ordinary grass or weeds. He then sought divination as was customary with the people in the olden days and God spoke to him that the wild saplings were his gift to mankind and if grown properly will produce grains and after cooking the end product can be eaten to sate hunger. The man was also reminded that a portion of the harvest should be kept for the next season to cultivate and this rotation or cycle will have to continue in order to sustain the same for generations to come.

Time rolled on and after a year of hard labour, the boy was able to establish himself independently. On one occasion while he was going towards his paddy field, he saw a woman sitting by the side of his plantation area. She enquired if that place belonged to him which he replied in the affirmative. She then asked him about his parents and his family to which he found it very difficult to explain, but told her that he had none and that he was reared up by the tiger and that the tiger had taught him how to cultivate and grow paddy. The woman was very surprised with this story of his and requested him to tell her more about it. He then narrated the entire story to her and how at the time of his birth his mother left him to his mercies and if it had not been by the kindness of the tigress he would not have survive. As he was narrating his life history to her, tears rolled down from the woman's eyes as she knew that, the boy was none other than her son, whom she had left behind on that fateful day. She could not control herself anymore and hugged the boy. The outpouring of her heart filled the ambience of that place with cries of joy and laughter as she had found her long lost son.

Both mother and son decided to live together and very soon the people in the neighbouring areas also came to know about it. Apart from the reunion between the mother and the son, the people were also very curious about the kind of work that the boy was carrying out as it was very new to the people of that area. They enquired about his occupation and he informed them about the knowledge that was imparted to him by the tiger and that if the people of that area would carry out this occupation they will never go hungry. This information that was provided by him made the people curious all the more and they decided to go to his place of work and to witness the same first hand. They were very much impressed by the neatness of his work and the manner in which the terraces were worked out. They even tried the rice which he had prepared and requested him to teach them, so that they could also learn the trade and that they could support their family and people. The boy taught the villagers about the technique of rice cultivation and with time the people also started to call him *U Stiar Hali,* and with that the cultivation of rice came to stay in this part of the world.

———————

(This information is being provided by Ms. Betty Laloo, Research Scholar, Department of Culture and Creative Studies, NEHU, Shillong).

12

The Legend of *U* Sajar Nangli

The story of U Sajar Nangli goes back to the time of the Jaintia king Borkuhain II (1731-1770). The legend provides that U Sajar Nangli was a *Lyngskor* (deputy) of Sutnga and was a resident of Ralliang and belonged to the Chadap clan[1]. He was a very important and powerful man during those days. U Sajar Nangli got married and had two children a son and a daughter and they grew up in the tradition of the *Pnar* people.

Trouble started in his relationship with the king when, the latter who by this time was already under the influence of Hinduism, was looking for a bride to be his queen. His messengers travelled across the kingdom to get a beautiful girl for the king, but he was not satisfied. Later he got the message about the beauty of the daughter of U Sajar Nangli and was determined to get married to her only. On one fine day the people of the king went to Ralliang to appraise U Sajar Nangli about the true intention of the king. U Sajar Nangli on hearing about the same, with politeness informed the messengers that he would not like to give his daughter in marriage to the king because he does not want that she and her children would follow the Hindu religion, but instead wanted them to stick to the traditional religion of the *Pnars*. The oral tradition claims that in reply U Sajar must have said these words: "*Nga da thung oo du chi tre u soh, tae kamon u e oo cha ki bru*" in other words meaning "I have planted only one fruit tree, how can I give it to others". Today there are a number of interpretations among the scholars about the true meaning of this statement, if it was ever being made by U Sajar Nangli. When the intentions of U Sajar was made clear to the messengers of the king, the latter then started making plans of kidnapping the daughter of U Sajar.

It was during this time that the Jaintia king announced that there would be a competition among his ministers and courtiers along with their followers to dig ponds and lakes all over the Jaintia kingdom. It was also said that the winners would be treated handsomely in the royal capital by the king himself. Along with the others officials of the king U Sajar Nangli also took part in the competition and in one single day he along with his people were able to dig three lakes and they are *Ka Bir Ympa Masi, Ka Mynkoi Tok* and *Ka Um Tisong* in the outskirts of the Nartiang village. It was clear that U Sajar and his followers were the winners in this competition, the king then invited them to the capital *i.e.* Nartiang, the place being the summer capital of the Jaintia king, and treated them properly. The king also informed his people to provide them with the best food that was available and to treat U Sajar and his men well. It was said that when it was time for them to have their food, the king's men started to serve Sajar and his men only with the leftovers. This had created a row and U Sajar prevailed over his people to finish their food and to proceed home. It was an insult to him and his people. It is not clear whether the same acts were carried out without the king's notice or whether it was done on his order. It was while they were getting ready to go home that the ministers of the king and his councillors requested him to stay for another night. But Sajar was adamant. The story goes that Sajar had already manage to get information from an old woman from a village called Mukhla, about the true intentions of the king and his men. They wanted Sajar to stay for another night so that they can kill him when everyone was asleep.

On reaching Ralliang, U Sajar Nangli came to know about the intentions of kidnapping his daughter and immediately called for a Durbar. It was held at a place called Phlong Khlieh Chnong and he spoke to the people informing them about the irreconcilable differences that he had with the king and how he was treated at the royal capital and their intention of killing him. U Sajar Nangli knew that his differences with the king would lead to an intensive warfare, causing a civil war among the people. He also knew about the outcome of such a disaster and wanted to avoid the tragedy by moving out of that area. With a heavy heart he left his native place, but he was not alone. His people followed him, and along with his family they started to move to the west. U Sajar and his men stopped by at Thadlaskein to take some rest. In order to commemorate the events and also for the others to know that he had passed by that area, he told his men to dig the place with the end of their bows. By the time his people passed by that area a huge lake had taken shape. This lake is being fed by perennial underground springs. The people of Raid Mukhla who continue to offer sacrifice near the lake are revering the lake till today. This lake is 56Kms from Shillong of the National Highway-44 and is considered to be a beautiful tourist spot.

U Sajar and his men kept on moving towards the west and some of his followers started residing in the Bhoi area. Interestingly, till today the people in the Bhoi region would still carry out their age-old ritual and ceremony called *Ka niam U Sajar*, perhaps to show their reverence to this leader. To this there is another tradition, which runs that U Sajar Nangli and his people moved towards the Kupli river. If this were true then it would mean that U Sajar Nangli and his followers moved towards the east. The Legend goes on to say that U Sajar Nangli moved further to the east and then entered Burma or Myanmar. Speculations are still ripe about the route of migration of U Sajar, and until further research work is being carried out it will continue to remain a matter of mystery to one and all.

NOTE

1. Baily Passah, "U Sajar Nangli Shadap" in S. Quotient Sumer, Panaliar Jowai Yungwalieh Souvenir, 1991, p.49.

13
Snikur's Wrath

Ummulong as a place has some stories to offer in one way or the other. In the outskirt of Ummulong village lays another small hamlet which is known to many as Snikur. The place got its name after the small stream which runs through it. The tradition of the people would refer to an old man who died in the village and his spirit instead of wandering elsewhere resided in the stream.

Like many other oral traditions this particular tradition is silent on the matter relating to the death of the old man and how his spirit came to be attached to the particular stream, Snikur. It was because of this kind of connection that the old man was generally referred to as U Tymmen Snikur (*tymmen* would mean old).

The spirit which hunts the river was regarded to be a malicious one and, therefore, would create a number of problems for the people. The most common of all the problems was sickness and this would be accompanied by high fever. In most cases a priest would be called to carry out a ritual where the sacrifice of a white rooster without any spots will have to be carried out. Prior to the sacrifice the priest will have to go to the stream and would seek for a, supposed to be dialogue with the old man's spirit. As a sign of an understanding between the priest and the spirit a blade of grass would be taken and a knot is being tied to the same. While carrying out the sacrifice, the priest will have to invoke the name of the old man, after which, his children and grand children would be invited for a feast which is being arranged for them. After all these have been carried out the person who is ailing is supposedly cured miraculously.

14

Some Notes on *Ka* Sara Dkhar

Ka Sara Dkhar was a woman who was residing at Khap Iaba in Chyrmang Village. She was married to U Pda Talang and had three children Ka Dhon, Ka Kupli and U Kieng. They had to struggle for every day existence and U Pda had always dreamt of having his own paddy field and some seedlings so as to sustain himself and his family. Interestingly, the task of growing rice in the village was monopolised by the people belonging to the Shylla Clan and they were not very keen on giving seedlings to anyone.

Pda Talang was aware about the intention of the Shylla clan but nevertheless he decided to go and to request them about the same. He was dressed up in a very rugged manner and left his long hair untied. This he did intentionally. On reaching the house of the Shylla clan he talked to them about his desire and they gave him grains that were being fried, their motives being that when he would cultivate those grains they will not grow. But then the torn and dirty clothes of U Pda had taken along with them some grains as they got stuck to his clothes. Some got stuck to his long hair too. His plan worked. He then took those few grains and cultivated them in his field. The members of the Shylla clan were furious on knowing about the deeds of U Pda and made plans to eliminate him. They caught him near a stream and killed him with a dao. This place is known till today as *Ka Wah Pam Pda* or the place where Pda was cut to pieces.

In the place where he was secretly buried a tree grew and legends has it that the tree talked to the members of the Pda clan, narrating about the entire events and how the members of the Shylla clan killed Pda Talang. Meanwhile when her husband went missing, Ka Sara Dkhar had to struggle to look after her three children, but her various activities of finding a livelihood for herself and her children would be ruined by the members of the Shylla clan. A time came when she was having problems of even getting drinking water. Getting fed up with these atrocities committed on her, she decided to go to Jaintiapur alone, leaving her children in a valley near a stream and told them to wait for her and not to go anywhere till she returns.

While she was away a man from Rymbai village came over to that particular place where the children were playing for fishing. He was attracted by the presence of the children in that place and somehow managed to coax them and took the sisters with him to his native village. When Sara came back in the evening she saw that her daughters were missing. She started looking frantically everywhere while her son, U Kieng wandered all over and finally landed in a place called *Wah Khah* or a stream full of reeds and eventually turned into a water serpent *U Thlen Um.*

Ka Sara managed to get information about the man from Rymbai who had taken her two daughters and went up to him to ask for them. The man heard her story but agreed to return only one of the sisters. Having no other alternative, Sara took the elder sister Ka Dhon and left Kupli with the man.

U Kieng who was residing in the stream was ill-treated by the people of that area, who would throw all rubbish into the stream and he moved away to another area called Wah Lapsep. He then called a meeting of all the water serpent of that place and they came to a resolution that since they were ill-treated by the people of that area, they would cross path with the people in their daily journey and who ever stumbles or falls would be inflicted with sickness by them. This would necessitate them to propitiate and to offer sacrifices to the *thlen* and in this way they can infuse fear and gain respect. This was carried out accordingly and these sacrifices that were offered by the people were a constant source of supply of food for them. Since U Kieng was regarded as the chief of the Thlen Um, a piglet was offered to him in the sacrifice.

This narrative ends abruptly without giving any details about Ka Sara and her other daughter Ka Dhon, much so with the case of Ka Kupli.

The people were also reminded that it would be a sacrilege for any members of the Talang clan to talk or to come into any kind of union with the Shylla clan. This tree which grew on the grave of U Pda is locally known as *U Dein Lein and* is being considered sacred by the people of the Talang clan and is a taboo to cut this particular tree.

15
Stone Cows

The story about the stone cows gives some hints to the kind of veneration that the people of a particular area would give to the said animal, so as to elevate them to the position of the sacred and referring to them as *Ki Blai* or gods.

Not very far from Nartiang and in between the villages of Mudop and Mynsngat, there lived a man who uses to shy away from his duties. The main occupation of the people in this area was rice cultivation, and because of his laid-back nature he was always running short of time. On one occasion when it was time for cultivation he took his cows to the field and started ploughing his land getting it ready before the first monsoon shower. Since he was already behind schedule he worked continuously without a break. He failed to show concern to his animals and put a lot of pressure and whipping them at times.

This continued for a few days till the cattle could bear it no more and ran away from the field along with the plough. The man gave chase but he was no match for them. The oral tradition offers that it was because of this continuous running and stress that they turned into stone and the man who came running after them on reaching near them also turned likewise.

The people of that area would refer to these stones as *Ki Masi Blai* or the reincarnation of the gods on these stones and a lot of respect was attached to them. This story by itself is a farfetched one but the implication of the same can be reinterpreted.

This place in the very first instance is close to Nartiang and the influence of Hinduism must have had a strong sway on the outline of the story. The selection of the animals, that is the cow, and not any other animal would gain in its importance from this angle, so as to venerate it and raise it to the position of being sacred.

The narrative also tries to throw light on the occupation of the people and the description of the cows, the use of the plough would talk much about wet rice cultivation in Jaintia Hills and that it was common with the people.

The oral tradition of the people also offers that this particular incident had taken place in the distant past and fixing the time frame would be a difficult task. This expression also serve to explain as to how with time the shape of the cows have been deformed and would resemble a pair of stone but close enough to put the plough on them.

One thing which is interesting about the stones is that though they have been elevated to the position of the gods, yet no sacrifices were being offered to them and no rituals were being carried out. If there is anything more that this story has to offer it goes in the form of a moral lesson, where the man out of his laziness and lack of respect to what he was doing found himself wanting in time, and in his mad rush to accomplish his task, which was already too late, lost everything and turned to stone instead.

16

Love and Loving gone Bad: The Case of a Jaintia Damsel

This story refers to the time when the *Syiems* of Malangniang were still strong and powerful and would exert their influences even in the border regions of the Jaintia kingdom, especially in the areas in and around Mookyndur, which was the seat of their power. This tradition offers that in those days the rulers would behave like autocrats and their rule is law. There would be no one who would dare to defy their authority and whatever that was pronounced by them will have to be obeyed and accepted. One such instance would relate to marriage. The king would pick up any girl that he wishes to marry and the consent of the other would not even be sought. If needed be the king may even resort to kidnapping in order to fulfill his wants and desires.

This narrative talk about one particular king of Madur-maskut whose name and time frame cannot be established, but the story is being narrated till today in the Jaintia hills. It was said that this particular king happens to come across one beautiful Jaintia girl and fell in love with her right away. It was love at first sight as many would acknowledge and wanted to marry her. The king then decided to complete the formalities by going to her mother and ask for her hand since her father had passed away. She was raised and groom by her mother and was the only child in the family.

One fine day the king showed at their doorstep and explained about his intention and desire of marrying the girl. Instead of being thrilled by the proposal, the young girl politely refused to comply stating that the king does not belong to her community and that she did not want to leave her mother as she was growing old. She pleaded with her mother not to get her married to the king.

The king on hearing this was taken aback but also assured the girl that he would send his servants to look and to take care after her mother but this promise failed to convince her. The mother, on the other hand, was afraid of losing her daughter but on second thought she knew that her daughter would have a secured life if she would get married to the king. The mother was confused with this turn of events. The king then offered the mother with gold and ornaments to which she accepted and tried to persuade her daughter to agree to the proposal. When the girl refused to listen to her mother, the latter allowed the king to take her away. The girl felt hurt and betrayed at the decision of her mother. The king then took her forcibly into his kingdom.

On reaching the territory of Madur–maskut, the king gave the girl a chance again to think about his proposal as he did not want to rush her in getting married to him. The king wanted to win her love before taking the final step. The girl was showered with the best possible things that one can ask for. She was treated with honour and dignity and there were servants to assist her at every step that she took. The king thought that this would be the best way of winning her love. Time went by and the king was also getting restless as the girl was not showing any signs of changing her opinion from her previous stand. His daily activities got affected as he could not help but keep on thinking about the girl. This created a lot of restlessness within him. He was anxious about her decision and the thought of losing her scared him. He realised that he was enslaved by the beauty of the girl.

After many days the girl walk up to the king one morning and requested him to build an open platform for her, to be raised high enough from the ground so that she cold enjoy the beauty of the surrounding area and to relax at times. The king was happy with the request and ordered immediately for the construction of the platform according to her wishes and specifications. A few days later she requested the king again for a golden necklace, a bracelet and the traditional Jaintia dress so that she could wear them at the time when she would be visiting the place. All her demands were met in good earnest by the king and hours were rolling into days. After a gap of few days the girl then asked for a basket to put her betel nut and leaf along with a knife made out of gold. Her final request was for a pair of pigeons to which the king provided without any kind of hesitation.

By this time the girl got whatever she wanted and on one fine day she dressed up herself with all the goods that were showered on her and requested the king to join her in that raised platform. She informed the king about her gratefulness for whatever he had done for her but then told him that she had one last wish from him. The king could not wait as to what it could be, as he was determined to please her. She then told him to get her a cup of drinking water as she would like to have the same from his own hands. The king immediately left the place and climb down the raised platform to get the water for his beloved. The girl in the meantime set free the pigeons and killed herself with the golden knife that was presented to her by the king himself. The pigeons flew passed the house of the girl and when her mother saw the pigeons she knew that something bad must have happened to her daughter as this was something that she had told her mother before the king took her away.

The king who came after sometime with the cup of water could not believe on what he saw. His beloved was in a pool of blood and that she had killed herself with the knife that he had presented to her. The king wept miserably and realised his mistake, for forcibly taking the girl and trying to get married to her without her consent.

News of this incident travelled far and wide and many came to know about this very sad incident. From that time onwards it had alarmed the men folk, especially those from the ruling class, not to use force in the process of selecting their wives and stress was given on the element of consent, which can be seen till the present day.

This incident was a sad reminder as to what can happen if women are taken for granted and that love speaks a different language all together.

17

Stories from Chyrmang

Chyrmang is a village in Jaintia Hills which is only around five kilometers from Jowai. This village is located on a hilltop and would give a panoramic view of the Syntu Ksiar valley. The inhabitants of this village were mainly the followers of the indigenous religion of the Jaintias, *Ka Niam Tre*.

One of the story which revolves around this village talks about *U Rakaw*, a reference to a large piece of stone which was erected by the elders of the village in the distant past. The stone is located inside the sacred grove which is locally referred to as *Khlo Blai* or forest of the gods. It is a taboo to hunt animals, to cut down trees or to show any kind of disrespect inside the sacred grove.

Tradition has it that this stone not only acts as a sentinel in the olden days but would also warn the villagers of the impending danger, in case of an attack from the enemy camp or whenever there is any kind of encroachment by others on their territory. This extra ordinary power of the stone was kept as a secret but it is something which became very difficult to uphold. News about the power of the monolith soon spread to the neighbouring villages too. This attracted the attention of others and attempts were being made from time to time to steal the monolith by people from outside the village. Every time they would dig and try to lift the stone, the same would sink deeper into the ground, making it difficult to steal and carry. As a result of this only a small portion of the stone can be seen today as the major part of it had gone underneath the earth.

The stone and the place are considered to be sacred today where the villager would perform a sacrifice every year as a mark of respect. In this sacrifice a cock, seven betel leaves and a betel nut along with the leaves of a particular shrub locally known as *sla lamet* would be offered.

18

Story of Birds and Animals

People living in the southern part of Jaintia hills have their own tradition as to how the animals and birds came into being. According to this narrative the elephant was initially a tiny creature and was missed or overlooked by both man and beast. In one occasion a festival was held where the animals and birds were involved in a lot of merrymaking, when someone pointed out that there was no water to drink and the elephant was requested to attend to the task in providing the same.

As part of the merrymaking, sacrifices were also made and while the elephant was busy carrying water the sacrificial meat was eaten and distributed. If room for introspection and interpretation is allowed one cannot miss the fallacies in these kind of oral narrative, where in the assembly of birds and animals meat was served to be consumed. Perhaps the narrator had failed to fathom this aspect or it could be otherwise, as a deliberate attempt to portray the picture of the animal kingdom. But whatever might be the case, the story goes on to reveal that when the tiny elephant came over with water, the others realised that nothing was kept aside for the elephant.

This made the animals and birds present there to think of doing something fast and they decided to give a bit of their own share to the elephant. Very soon the elephant found itself with the largest share of all and after taking the same it grew into an enormous size. This story also has a reference to a small bird the wren, who gave away almost everything except its heart. It is said that because of this act of kindness where it gave away almost everything, it remained as the smallest among birds till this day.

To this story, there is another version which states that every animal and bird was created by God out of different substances. But then after doing the same he found out that the quantity of leftovers was too much to be wasted. He decided to use the various substances to make one last animal and that was the elephant. The people do believe that the flesh of the elephant is made up of a great number of portions, each with a different taste, which is the taste of different animals and birds.

19

Story of the Earth

The story of the earth has a lot to do with the kind of geographical conditions that the Jaintia territory is being blessed with, not forgetting the rocky and the sterile character of the hilly tracts. In contrast to the Jaintia scenario the fertile conditions that are prevailing in the plain region had made the people to come up with a story and the same is very common with the people of Amwi in the war areas of Jaintia Hills.

The narrative starts by referring to the earth as an entirely rocky area and that soil was not available for any kind, neither for cultivation nor habitation. In order to tackle this problem, the gods thought of bringing soil form the interior of the earth and spread it over the barren surface. The task for bringing soil from inside the earth's crust was given to the white ants. The act of distributing the same to the different parts of the earth was given to two small birds locally known as U Simsong or the Redstart and Ka Syiem-madiah[1].

Separate areas were assigned to these two birds. The former was to work in the hilly region and the latter was to do her job in the plain areas. U Simsong happened to be a vain and pleasure seeking creature and in the course of his flight came upon a group of gamblers who lured him into their game[2]. The bird by this time who was already bored with the task that was assigned to him, agreed to join the group. But the game did not go in his favour and after losing repeatedly, he staked the soil that he was carrying on a last desperate throw, but lost even that. In panic he fled away and hid himself among the rocks of a stream.

Ka Syiem-Madiah on the other hand was just the opposite of U Simsong. She toiled hard and completed her work. She then went to report to the gods who in return were surprise to see that the hills were largely bare and that U Simsong was nowhere to be seen. The Syiem–madiah because of her good nature, took out whatever soil that was there in her bag and spread it over, but it could barely cover the rocks.

The gods on the other hand were disappointed with the activities of U Simsong and as a punishment for his misdeeds, placed ashes on his forehead and condemned him to remain forever among the rocks in the riverbeds.

This story seeks to explain as to why the hills are completely barren and why the Redstart spends its days among the rocks of streams and rivers.[3]

NOTES

1. I.M. Simon, "Some common Themes in Khasi Oral Literature" in Soumen Sen, Folklore in North East India, p. 160.
2. *Ibid.*
3. *Ibid.*

20

The Case of the Jaintias with Madur-Maskut

While trying to understand the case of the 'Divine Descent' of the Jaintia kings a reference is being made to a chief by the name of U Ksinor Saitsner or the intestine washer residing at Borkhat and having super natural powers, the secret of which lies in his ability to take out his intestines, wash them and keeping it back inside his body. Mention is also being made about a young boy by the name of U Markusain, who was very mischievous and was sent by his parents to stay with U Ksinor to correct his attitude and how the young boy was reformed. The story also reveals that how one day, Murkusain was able to spot the chief taking out his intestine and washing them. As a result of this the chief died on the spot and Markusain was able to incorporate the territory of the chief with that of Sutnga and became the undisputed ruler over a huge area. The narrative is, however, silent as to how the territory of Sutnga came under the control of Markusain and that he managed to bring Borkhat too under his fold without any kind of reaction from the people.

What appears to be striking at this point of time and especially when we refer to the case of the rulers of Malyngiang was the reference to another ruler who had the same feat as that of Ksinor of Borkhat. The oral tradition talks about one ruler of Madur-Maskut by the name of U Kyllong Raja and has magical power in his intestines[1]. This has made him invincible in many of his quest to acquire more land and territories for himself. The place Madur is near Moodymmai and Maskut is in the Sung Valley[2], which is located in between Jowai and Shillong on the National Highway No. 44. This would also mean that in the olden days this principality ruled by the Malngiang was in between the kingdom of Shillong and that of the Jaintias. It was because of this close proximity that the war between Madur-Maskut and Shillong or for that matter between Jaintia and the former was very much in the offing. The wars were fought mainly to extend one's own territory and this in so many cases would spark out because of the kind of encroachment that was carried out on the other. It was held that at one point of time the kingdom of Madur-Maskut extended itself over the whole of the Khasi and Jaintia principalities, extending up to Damera in Nowgong in the North and down to Kuwain in Sylhet in the South[3]. Since the ruling authority came from the Malngiang clan[4], the rulers were known as the *Syiem* or kings of Malyngiang. Their capital and fort was located on top of a hill where at present is adjacent to a place called *Iaw Dai Ja* or rice market.

Another version has it that the progenitor of the Malngiang clan was a person with strange magical powers, whereby in the night he would transform himself into a pig and during the day into a human being[5] and would lead an ordinary life. The chronological time frame that was given to these rulers was in the 15th and 16th centuries and strong references is also being made about them during the time of Majha Kuhain or Markuhain or Markusain who ruled the Jaintia kingdom from 1516 to1532 A.D. This kind of reference came about because of the kind of enmity that was there between Madur-Maskut and Jaintia over the acquisition of territories. That time, Madur-Maskut was ruled by U Mailong Raja and Jaintia was under Majha Kuhain. On one occasion, U Mailong Raja invited the Jaintia king and all his deputy to his kingdom in order to work out ways and means as to how the boundary disputes between the two could be solved. The Jaintia king on learning about the invitation from his adversary thought that it was an opportune time to go to Madur-Maskut and to settle the matter with U Mailong Raja once and for all. As it was customary to call a durbar before any such great event, the Jaintia king also did the same and a lot of deliberation took place where many of the elders tried to analyse the purpose and intention of such an invitation. The elders from Jowai agreed to go but hinted that the king must remain within the kingdom and to grant permission to his Dalois to go and address the problem. They were still apprehensive about the true intention of the king of Madur-Maskut. To this effect, they got the king's officers U Kaia Bandari and Katha Sohwata[6] to dressed up like a king and a member of the royal family respectively. Two women were also added to the list, as members of the royal family and they were Ka Lapang Rani and Ka Dhai.

It was in the place called *Iaw dai Ja*, that U Mailong Raja received them and a grand feast was arranged as a mark of respect for accepting the invitation and also to mark the occasion. Proper care was taken to see that the Jaintia king and his people were getting the best food and drinks. The local liquor that was given to them was known as *IaIaroh*[7] and it was just a matter of time when a good number of the Jaintia delegations were in an inebriated condition. When U Mailong Raja saw that the Jaintias were not even in a position to stand up, he gave the order to execute all of them. The people of Madur-Maskut came out of their hiding and massacred their victims who were now helpless and reeling under the influence of alcohol. The death toll was around 460 to 480[8]. The news about this massacre reached Jowai and the king Majha Kuhain was informed about the treachery that was committed by the king of Madur-Maskut. The people of Jowai were enraged and the thought of taking revenge was already in the air.

Meanwhile the king of Madur-Maskut was already making plans for the annexation of the Jaintia territory. Preparation for the same was gaining ground at Madur-Maskut and the king ordered that there should be a celebration and merrymaking of nine days within the kingdom before their final departure towards the Jaintia territory. Accordingly, there was a lot of merrymaking. The Jaintias were able to infiltrate the Madur-Maskut camp with the help of the two slaves of the king of Madur-Maskut, U Tonkha Laloo and U Buitkha Laloo who were originally from Jowai, and were the favourite slaves of the king U Mailong Raja. The duos were in Jowai when the incident at Madur-Maskut happened. The task of taking revenge was spearheaded by the people of Jowai.

After working out the details with the two slaves who would be returning to the King Mailong Raja, the people of Jowai selected around sixty well-bodied men to carry out this task. On the last day of the celebration as was planned, and with the help from the two slaves, the sixty men from Jowai managed to enter the fort of Madur-Maskut and killed all the people who were sleeping and burnt the fort. U Mailong Raja was fortunate enough not to be there in the fort on that night as he had gone to his wife place at Madur. On hearing the cries and seeing the fire that was burning which was coming from the direction of his fort, he took out his horse and raced towards his capital. The Jaintias who had already knew about his movements earlier, decided to waylay him at Moodymmai junction. The raja was killed at that place and his head was severed from his body. The revenge was taken. The skull was taken to Jaintiapur and a replica of the same, which was made out of silver was still there in Jaintiapur[9] till the days of Ram Sing II (1790-1832). The massacre was carried out to perfection, as it was planned meticulously. It was said that only a few managed to escape in the dark. When everything was over these few came back and started residing in a place which is presently known as Maskut.

Tradition has it that the collection of stones at *Iaw Dai Ja* was erected by the people of Jowai as a mark of remembrance for those Jaintias who were killed by treachery on the orders of U Mailong Raja.

What is perplexing about the case of the Malaniang Rajas' is the anonymity of its rulers. One fails to understand as to how a powerful kingdom, as claimed by local writers, which have extended its territories over the Khasis and the Jaintias and furthermore into the plains of Sylhet would escape the attention of the other neighbouring powers. There are no accounts about the lists of rulers either in the form of writings or oral traditions and whatever that is available is also conflicting. This can be seen in the subsequent discussion.

After the assassination of U Mailong Raja, another ruler who had made his mark in Madur-Maskut was U Kyllong Raja[10]. There are many things that were attributed to this ruler. He was not only regarded as a brave king but was also having super natural powers. He was an ambitious ruler and it was because of his desire to extend the territories of his kingdom that he often ended in a war with that of the ruler of Shillong or that of Jaintia. In one occasion, in his fight with the Jaintias, the Jaintia ruler was able to kill him and cut him into pieces, but it was said that in the following day they saw him moving all over his dominion. Tradition has it that since U Kyllong Raja was having this super natural power stored in his intestine, it would be very difficult for anyone to kill him, until and unless his secret powers is taken and destroyed completely. In the writings of H. Elias, *Ki Khanatang U Barim,* he referred to this ruler with extra ordinary powers as U Niang Raja. It is still difficult to ascertain at this point that whether they are the one and same person. Both H. Elias and Homiwell Lyngdoh had written their books in 1937 and 1938 respectively and it is difficult to imagine at the same time as to how they could stray from their sources or how their sources could be so different. At this point, if reference to Ksinor the chief of Borkhat, who was also known as an intestine washer, is to be included, it would add another dimension to the already perplexing problem.

Though a different name is being attributed yet the story line especially in the case of U Kyllong and U Niang Raja remains the same, without diluting the content. In both the cases, the story would revolve around the role played by a young and beautiful damsel from Nongbah and under the instruction of the Jaintia king was willing to sacrifice her chastity in order to perform her duty and service to the king and her people. She managed to charm the actor and very soon got married to him. She managed to get the details from him about his secret power and when the same was revealed to her, she was able to communicate the same to her ruler and in the process the Malaniang ruler was finally eliminated. The story offers that on the day when the king went to his secret stream to have his bath, as usual he would take out his intestines and washed them. At that time the Jaintias who were already spying on him sent one man from Raliang village to attack him. This he did with great accuracy and his next move was to cut the intestines into pieces. This was not enough as the mere cutting of the same will help the Raja to recover himself. The next step that was taken was to feed the intestines to the dogs and with this the power of the Raja was totally destroyed. Here is a strong case of love and betrayal and also talks much about the kind of diplomacy that was worked out by the Jaintia ruler.

After this incident the Jaintias took over the territories of the Malaniang rulers and incorporated it within their fold. Homiwell Lyngdoh had put a rough estimate that the kingdom of Madur-Maskut was eclipsed by the Jaintias in the year 1651, the year when Josamanta Roy (1647-1660) was ruling over the Jaintia kingdom. Right from that time the rulers and subjects of the Madur-Maskut kingdom had scattered throughout the Khasi and Jaintia territory.

NOTES

1. Homiwell Lyngdoh, Ki Syiem Khasi Bad Synteng, p. 3.
2. Interview with Ms. Betty Laloo, Research Scholar, Department of Center for Cultural Studies, NEHU; and Mr. Edmond Lamare, Research Assistance, Department of History,NEHU, on 27 July 2010.
3. *Ibid.*
4. *Ibid.*
5. H. Elias, Ki Khanatang U Barim, p.173.
6. Homiwell Lyngdoh, op.cit., p.37.
7. *Ibid.*
8. *Ibid.*
9. *Ibid.*
10. *Ibid*

21

The Five Clans of Mukhla Village

An interesting oral tradition which is in circulation in the Mukhla village would relate to the origin of the five clans namely the Suna, Shylla, war, Lyngdoh and Sari. The story traces itself to the time of U Sajar Nangli and his migration from the winter capital of the Jaintia King, *i.e.*, Jaintiapur and then to the upland region. Along with the others who came with him, Sajar nangli was accompanied by the three sisters Ka Na, Ka Doh and Ka La.

On reaching Mukhla village they planned to settle there and not to go further with U Sajar Nangli and his people. Tradition has it that form the three sisters the Suna, the Lyngdoh and the Shylla clan emerged. In the mean time a woman form the southern part of Jaintia hills which is being referred to as the War area came to settle in Mukhla village. The people identified her with the area she came from and in the process she was called War and was the progenitor of the War clan.

While these settlements were taking place in the village another Muslim girl came over as a refugee. The oral narrative provides that she was in her early teens when she came over to Mukhla village. Doh took sympathy towards this young girl and adopted her as her own child. Since no one knew her name and that she was wearing a Sari, this name came to be attached to her and was identified as Sari, by Ka Doh and the others. This was not taken very well by the other two sisters who have always considered her to be an outsider. Things did not go very well with the young girl as she was subjected to torture and other kinds of punishment by the other two sisters. On one occasion they wanted to get rid of her and took her to Nartiang, a nearby village, with the intention of leaving her there. Doh who came to know about the evil intention of her sisters managed to free the girl from their clutches and brought her back home. This was followed by a stern warning to her other two sisters and refrained them from doing anything bad to the girl.

With time Sari grew up to be a beautiful lady and later was married to a man from the Lyngdoh clan. Her children took her name and in this manner the sari clan came into existence in the Mukhla village.

Mukhla village has another story to offer. It was said that after U Sajar Nangli and his followers had dug the Thadlaskein Lake, before leaving Mukhla village, they had a meeting at a place called *Moo-Pyniein* or standing stones, to decide on their onward journey. This place was so called because of the large collection of stones which are believed to be erected by the followers of U Sajar. The people of the Raid Mukhla would still pay homage to this place.

In the case of the Moo-Daloi Pator, it was considered to be a meeting place between the people of the Raid Mukhla and the Daloi of Nongbah. Prior to the coming of the Europeans to the hill and the English in particular, the people of the Raid Mukhla would perform rituals and ceremonial dances in the presence of the Daloi of Nongbah at Thadlaskein Lake in order to pay respect to U Sajar Nangli. The performance used to be held once in three years. The customary practice was such that the people of the Raid Mukhla would wait for the Daloi of Nongbah and his followers at a place called Moo-Daloi-Pator. On reaching the place the Daloi of Nongbah would distribute money shidamli (which is less than shinaia or one paisa) to the people of raid Mukhla and then would proceed to the lake along with the others from the raid, accompanied by singing and dancing.

22

The Incursion on Sylhet and the Fallout with Nongkrem

VERSION I

The relationship of the Jaintias with that of their immediate neighbours seems to depend a lot on the aphorism that there cannot be any peace if there is no war. Therefore, in many cases, war does serve as a means to buy peace, more so with those who were trying to exert their influence. This narrative between the people of Jowai and that of Nongkrem will in so many ways talk about the enmity that was there between the people and the kind of diplomacy that was carried out by seeking the help of another party in order to suit their interest. Though some of the events narrated would be a farfetched kind of a claim, yet the oral traditions of the people do talk about the role of the individuals, with a dash of historical developments intersperse in-between.

To refer to the exploits of U Kiang War, the historical developments that had taken place before he came into the picture have to be looked into. The spoken legends of the people would go back to the time of U Bor Kuhain II (1731-1770) when the Jaintias raided the plains of Sylhet and looted whatever that was there. Prior to this attack there was a general understanding that the spoils of the war would be distributed among those who had taken part in the war. This had given them that necessary drive to be ruthless in their pursuit and campaigns. The war was carried out effectively and on their returning back to Jaintiapur, they found out that U Kiang Lyngskor had hidden many of the riches, which he had brought along, for his own personal consumption. This was not taken very well by the other soldiers and the matter was brought to the notice of the king Bor Kuhain. Since U Kiang belongs to the hilly portion of the kingdom, the king could not carry out his judgement without consulting the people of the hills especially those of Nartiang and that of Jowai. It was decided that the king would send a bracelet along with the accusation charges to both the villages and accordingly they should give their verdict. This seems to be the kind of practice that was prevalent in the past as to how the people they should proclaim their judgement or pass their sentences.

Both the villages called for their durbars and they came with their own interpretation and more importantly, on the kind of punishment that was to be given to U Kiang Lyngskor. The people of Nartiang decided to sent back the bracelet to the king only after making a slight mark on it, indicating that the person concerned should not go unpunished but the punishment meted out to him should not be harsh either, and therefore, the slight mark. The people of Jowai on the other hand came to a resolution that the bracelet should be cut into two, thereby, indicating that the person concerned should be given the capital punishment, by beheading, as this was the practice that was prevalent during those days.

The time came for the king to announce his verdict and he proclaimed the decision that was taken by the people of Jowai, and accordingly U Kiang was to be executed. Before the execution U Kiang was asked if he had anything to say as a condemned man. He replied in the affirmative and said the following words: *"If I am a sinner, after the execution, my body will fall towards the west, but if it falls to the east then I am not a sinner and that I am accused of something that I have not done, but at the behest of my enemies."* Many were there to witness the execution and when the execution was carried out, much to the horror of all those present there, the headless corpse fell towards the east an indication that if his words were true, they have killed an innocent man. A few people from Jowai reacted very strongly to what had happened and were very much convinced that the decision to execute U Kiang was mainly carried out by some who were jealous of him. These considerations have divided the people into two camps, and the group which had sympathised with the case of U Kiang were convinced that those who were responsible for the same should not go unpunished. They also realised that by going to Jaintiapur and to get those who were responsible for spreading this rumour would amount to going against the king's men. They then decided to muster some support from the ruler of Khyriem (who were mainly referred to by the Jaintias as Khynriam) and with this thought in mind they went over to Nongkrem.

The rulers of Khyriem had always had their enmity towards the Jaintias and this had a lot to do with the manner in which the kingdom of Madur-Maskut was liquidated and the kind of kidnappings that was carried out in the border regions. Therefore, when a group of Jaintias came out with this idea of going to Jaintiapur and to attack those who were responsible for the killing of U Kiang Lyngskor, the ruler of Khyriem took this as an opportunity to take revenge on the Jaintia ruler. Preparation was made and the attack on Jaintiapur was on its way. The entire planning was kept a closed guarded secret and the Jaintia King Bor Kuhain II did not have the slightest idea that his capital was under threat.

The sudden attack that was carried out by the Khyriem Raja worked to his advantage and the Jaintia king along with his sister Gauri Kuari or Ka Kong Rimai as she was locally known, along with U Mon Kongor and U Le Ki Sing were kidnapped and taken to Nongkrem. All the four were kept in a small room the floor of which was covered with cow skin. This was done deliberately to taunt the king and his people about their faith and belief system. As for the Royal family it was a sacrilege to sit on a cow skin as they have already embraced Hinduism.

News about the kidnapping of the king with the members of the royal family soon reached Jowai and the plight of the Jaintia family in captivity spread like wild fire. The fact that no respect was shown to them by the ruler of Khyriem and that they were not able to eat or drink made the people of Jowai very angry. A meeting was called and it was resolved that the King and the others have to be freed from captivity as soon as possible. The people of Jowai started to work out a strategy and also realised that they will need the help of the rulers of Mylliem and that of Sohra should there arises a necessity to transport their rulers through these areas. Interestingly, the rulers of Mylliem and Sohra during that time were not in good terms with the ruler of Khyriem and they decided to help the people of Jowai. On the appointed day the people of Jowai went over to Nongkrem and stayed overnight. They were able to smuggled themselves into the house where the king and the others were kept and within no time they were able to whisked them away. The King and his sister were taken to Sohra, while U Mon Kongor and U Le Ki Sing were taken to Jowai.

On the next day, by the time when the people of Nongkrem had realised what had happened, the king and the others were already in their respective areas as planned. This had enraged the king of Khyriem and his men and they decided to go down to Jowai and to challenge them there. This tradition has it that the king of Khyriem who was responsible for the kidnapping of the Jaintia king and his sister was U Hain Manik[1]. Another source would refer to the ruler as U Ksan Shillong and U Bormanik Syiem of Khyriem[2] who had carried out the same. The tradition goes on to offer that when the latter two and their men reached Jowai, they accused the people of coming over like thieves and taking away their king. They also used a lot of foul language in the course of their accusation, which enraged the people to a very great extent. When things became unbearable any longer, a man by the name of Jata Syntan, took out his bow and arrow and shot at U Ksan Shillong. The arrow narrowly missed its target, but grazed U Ksan's turban. The intention was to hit him on the forehead and to kill him on the spot. U Ksan and his followers from Nongkrem decided to leave but with a warning that they will be back for revenge. On reaching Nongkrem, they called for a meeting and decided to go for war against the people of Jowai. They were also convinced that they have a reason to go for war as U Ksan was already attacked. The usual practice in the olden days was that, they needed a valid reason to go for a war and in this case the events that had taken place in Jowai was good enough for U Ksan and his men to declare war on them. As a result of this a large number of them marched towards Jowai and on reaching a place called Moo-li-ksoo, the people of Nongkrem decided to built a fort for themselves in anticipation of an attack. Moo-li-ksoo is a small hill which is also known as Pohkablai and is centrally located today. With the expansion of the place, the hill Moo-li-ksoo is being surrounded by four localities. In the north by Tpep-pele, in the east by Dulong, the south by Iongpiah and the west by Mission compound. Today this small hill is covered with tomb stone and some form of religious rituals is still being carried out especially the one relating to the bone intern ceremony. It was from this place that the people of Nongkrem would come out sometimes during the day and at times in the night to attack the people of Jowai.

Unlike in the case of the other villages, spread out in many parts of the Khasi and Jaintia territory, the people of Jowai had never fortified themselves and perhaps this was one reason as to why it was very easy for the people of Nongkrem to come all the way and to entrenched themselves almost in the heart of Jowai. A serious fight broke out but the soldiers from Nongkrem were finding it hard to break the resilience of the people of Jowai as they were ready to fight to the finish. It was in this struggle that a reference to one man by the name of U Kiang War becomes prominent. This man refused to remain confined within his own camp and decided to come out of his hiding and to face the people of Nongkrem by attacking them. He would carry a huge shield in order to protect himself from the arrows that were fired upon him. At times his shield would be very heavy because of the weight of the arrows and this would necessitate him to go back inside and to take out the arrows from his shield, and then to come out and fight again.

This assault on Jowai did not result into something concrete as far as the people of Nongkrem were concerned. They decided to spread their people in two directions. One group went towards Laturiem and were making a lot of noise as they were proceeding towards that place and they built a temporary camp for themselves there. The other group went towards Chyrmang in the still of the night and on reaching that place, killed all the people in the village and razed it to the ground by burning everything in it. The people of Jowai thought of moving towards Chyrmang village to help their fellow friends but were prevented by the Daloi U Mut, who prevailed upon them that Jowai has to be protected and should the others go, then it will be vulnerable to further attacks. U Kiang War disobeyed the order of the chief and proceeded towards Chyrmang village. Following him was U Kiang Gatphoh, another fighter from Jowai. They moved from the river Syntu Ksiar or the Golden Flower River and managed to reach Chyrmang village without any difficulty. The people of Nongkrem taking pride on their success resorted to drinking and not a single one of them was sober. U Kiang War and U Kiang Gatphoh, enraged by what the people of Nongkrem had done to the people of Chyrmang, killed every single one of them. Meanwhile in Jowai the Daloi was having a tough time to control his people as they all wanted revenge. He had to succumb to the pressure of the people and they all came out in the open and started attacking the people of Nongkrem who were hiding at Moo-li-ksoo. The people of Nongkrem had to retreat back from where they came but was given a responding chase by the people of Jowai. Reference is also being made about the role of U Tep Dkhar a brave and strong warrior from Nongkrem who was trying to protect his people by retreating last and that of U Pati Laloo from Jowai who was in hot pursuit. The clash between the two was inevitable and U Pati Laloo was able to attack U Tep with his sword and injured him from the leg. At that time U Pati Sohwata took out his sword and cut Tep into two from the waist portion. The people of Nongkrem on seeing this realised that their end would be near made haste about their escape and managed to enter their territory before being massacred by the people of Jowai.

When this serious struggle was going on in Jowai, the ruler of Sohra managed to take the Jaintia King Bor Kuhain II and his sister Ka Kong Rimai to Jaintiapur and on reaching the capital as a token of good will for all the help rendered by the king of Sohra, two important places, namely, Angajur and Fatehpur was given to the latter by the former. Right from that time these two places were under the control of the king of Sohra. All these happened before 1770.

VERSION II
THE KIDNAPPING STORY RETOLD: THE CASE OF BORKUHAIN II AND BOR SING OF MAWSMAI.

This narrative refers to the time when U Hain Sing was the king of Nongkrem and U Bor Sing was that of Mawsmai. The Jaintia ruler during that time was U Bor Kuhain II. In one occasion it was said that U Bor Kuhain and his sister Ka Kong Rimai left their kingdom to visit places in the neighbouring region, the territories of the khasi rulers. There is something amiss with this tradition right at the very beginning itself. It is difficult to comprehend as to how the head of the state would move along with his sister without any armed escort, considering the fact that U Bor Kuhain II was already regarded as an important ruler of the Jaintia kingdom. Furthermore, whether some form of a protocol was in existence, especially for a ruler to enter a different territory altogether, without the knowledge of the ruling authority beforehand is yet to be ascertained.

This tradition does not seek to give any clarification or better still does not intend to justify any of the assertions but has the following to offer. When the two reached Nongkrem, they visited the market place, as this was and still is, one of the important areas for the people of the region, to spend their time there. News of the presence of the Jaintia king and his sister in the market place of Nongkrem soon spread over and the king of Nongkrem U Han Sing, instead of welcoming them into his kingdom, gave an order that they should be arrested and to be kept in a pig sty, where the floor was covered with cow's skin[3]. This is something that the members of the royal family cannot do as it goes against their religion, for by this time the members of the royal family were already Hindu converts, belonging to the Sakta sect. In addition to this, it may be pointed out that the tantric cult had made inroads into the Jaintia kingdom where sacrifices would be offered to the goddess Kali based on the teachings of the religious text the Kalika Puran[4]. This would also explain as to why the Jaintias in the olden days, they would not slaughter cattle for food, but treat them with respect.

Bor Kuhain and his sister were kept in captivity, far from the reach of ordinary people and were guarded round the clock. No one was allowed to come near them or to talk to them. This went on for quiet sometime. Meanwhile in the Jaintia kingdom the story about the sudden disappearance of the king and his sister had spread far and wide and meetings were held from time to time to find a way out. It was in one of those grand durbars that a resolution was passed that the task of spying and searching for the king and his sister was assigned to Patiphas. According to the oral tradition, Patiphas was one of the courtiers of the king in Jaintiapur and was a man who was very witty and full of wisdom. He was advised to tour the khasi territories in disguise and to gather information about the whereabouts of the Jaintia king and his sister. He went to many parts of the khasi area and on reaching Nongkrem came to learn about the arrest that was made and the location where the prisoners was kept. Till this point he was not sure that whether the persons kept in captivity were the ones he was looking for, but this again had made him to be very curious all the more. After a lot of hard work and with his knack in getting out information from the people there, within a few days time, he came to know that the persons in the lock up were none other than the King and his sister themselves. At first he was very happy to know that they were still alive, but then the task that was there ahead of him, as to how to free them from captivity was an arduous one. He was very careful in his movements and in his mannerism, as he did not want to attract unnecessary attention. He tried to get help from within Nongkrem itself, but when this failed he turned his attention towards the ruler of Mawsmai, U Bor Sing, who during this time was a very influential king.

In his meeting with the king of Mawsmai, U Patiphas apart from the many things that was shared with U Bor Sing, he also assured him that half of the plain territories under the control of the Jaintia king would be given to him, provided he could secure the release of U Bor Kuhain from the hands of U Han Sing. The king of Mawsmai who had always had his eyes on the plains of the Jaintia kingdom was attracted by this proposal, but did not agree to the terms and conditions until it was re-assured by the Jaintia king himself. For this U Patiphas will have to work it out as to how to get the consent of the Jaintia king, so that the same could be conveyed to the king of Mawsmai. Keeping in mind the kind of security in which the Jaintia king was kept in U Bor Sing suggested that U Patiphas should dress up himself as a wandering fakir, who by mistake have landed himself in Nongkrem, and would be playing the sitar and singing songs in the Jaintiapur dialect so as to communicate to the king in that dialect and to give his consent. U Bor Sing also expected that the Jaintia king should give a strand of his hair as a mark to seal the agreement.

As was planned U Patiphas went to Nongkrem on a market day and started singing songs in the Jaintiapur dialect much to the amusement of many who did not understand what was being sung. The Jaintia king and his sister were also brought out from the place where they were kept and were made to listen to what the fakir had to say. It was an unexpected turn for Patiphas, as he did not expect that things would start working in his favour. With this the message was conveyed to the king and he plucked the strand of hair from his head and handed it over to the fakir so as to conclude the agreement with U Bor Sing. The fakir (U Patiphas) then left the place and proceeded towards Mawsmai to convey the message to U Bor Sing and about the consent of the Jaintia King to part with a portion of his plain territory if he could free him from the plight that he and his sister were in. U Bor Sing on learning about the details from U Patiphas, started to make arrangements and gave some money to U Patiphas with the instruction that he should proceed once again to Nongkrem and to carry out the same task as he had done earlier but by nightfall he should be able to treat the guards to country liquor and to make them drunk. The plan was executed accordingly. Patiphas then broke the lock of the main door and guided the king and his sister to safety. Arrangements were already made by U Bor Sing to have them carried away by men-carriers in the special cane chairs meant for the said purpose. U Han Sing the ruler of Nongkrem realised in the morning that the Jaintia king and his sister had managed to escape. He along with his men was in hot pursuit, to capture the Jaintia king once again. U Bor Sing who had already anticipated the confrontation with U Han Sing, had made all the necessary preparation in advance. A fight between the two parties broke out at river Umiam and U Han Sing soon realised that the fight was an unmatched one and that it was difficult for him to recapture U Bor Kuhain II as he along with the others had already cross the river. Attempts were made by U Han Sing and his men to fire arrows to the other side but to no effect. In despair they retreated back to Nongkrem. Except for this small hiccup at Umiam, U Bor Sing along with U Bor Kuhain II and his sister Ka Kong Rimai, including their followers reached Mawsmai without much problem.

After a few days of their stay at Mawsmai, U Bor Kuhain and his sister decided to leave for Jaintiapur and invited U Bor Sing to accompany them and also to see the places that was to be given to him as per the promises that was made. The latter also agreed to the proposal and took U Phiah Lyngskor, one of his deputies and a few servants to accompany them on their journey towards the plains. The route that they took was from Mawsmai to Nongjri and then to Lyngkhat and then finally to Angarjur Fatehpur which was the territory of the Jaintia king. They were in this place for a few days and many of the courtiers and deputies of the Jaintia king were also present to meet him. They stayed in this place for some time and then moved towards Jaintiapur.

Bor Sing's stay in Jaintiapur was made very comfortable by the Jaintia king, but then the latter avoided any kind of discussion on the issue as to how and when the territories would be handed over to the former. Attempts was made by U Bor Sing from time to time to get the Jaintia king involved in the discussion about the matter but U Bor Kuhain would avoid it on one pretext or the other. U Bor Sing could not understand this change of stand and attitude on the part of U Bor Kuhain but then thought that it could be because of work pressure and other problems related to the kingdom due to his long absence. These things went on for a few more days.

On one fine day U Bor Kuhain invited U Bor Sing for a boat ride and to go down towards Sylhet. U Bor Sing who did not suspect any foul play agreed to the same. Two different boats were arranged for the purpose and they slowly moved towards their proposed destination. Everything seems fine initially, but then Bor Sing realised the problem when he could not see the boat of U Bor Kuhain. He then told the boatmen to turn the boat and to go back to Jaintiapur, which they refused, but instead moved ahead to the fort of a big Zamindar in Sylhet by the name of Sujlar. By this time everything was clear to U Bor Sing that the Jaintia king had not only betrayed him but had kidnapped him, with the intention to imprison him in that particular fort in Sylhet. In the evening U Phiah Lyngskor was getting restless when he found out that U Bor Sing had not arrived, whereas U Bor Kuhain was already there in the palace. He tried to get information from the Jaintia king but to no avail. This made him to think that something bad must have happened to U Bor Sing and that U Bor Kuhain was instrumental for the same. When no news of U Bor Sing was coming through, U Phiah Lyngskor then went and approached U Potiphas and narrated the whole story to him. Without losing much time U Potiphas headed towards the palace to have a discussion with U Bor Kuhain, but the Jaintia king did not pay any heed. Things took a different turn altogether when the Jaintia king was adamant to release U Bor Sing. U Potiphas felt betrayed and let down by the action of the king and decided to send U Phiah Lyngskor back to Mawsmai. Immediately after the departure of U Phiah Lyngskor, U Potiphas revolted against the action of the king and the entire royal palace was in turmoil. The king tried to control the situation but things were getting worst with each passing day. The king then got Potiphas murdered in his sleep with the help of his own wife and some of his guards, by promising to pay the wife a huge sum of money.

The people of Mawsmai on coming to know about what had happened to their ruler from U Phiah Lyngskor decided that people should be sent in all directions to spy and to provide information about the whereabouts of their king. The attempts to get information and also to know that whether U Bor Sing was still alive or not, have always ended in failure. The tradition has it that three years have passed down the line and nothing substantial could be obtained. Then on one occasion a man from Wahlong by the name of U Kong Sing who use to frequent Sylhet very often provided the information about the man whom he saw when he was near the fort of Sujlar, who looked very similar to the one that the people of Mawsmai were looking for. This information created a lot of excitement among the people and decided that they should select the six clans of Mawsmai to go down to this fort and to confirm as to whether the information provided is accurate. U Kong Sing also agreed to accompany and to guide them till they have reached the place.

Initial attempts were made on one morning to make some kind of contact with the person whom they thought to be U Bor Sing, and this was made possible through a narrow channel of water that was running into the fort. Bor Sing who got up in the morning and went to have his wash near the channel saw a leaf that was tied to a thread and on bending saw a group of people who were showing signs to him. The group on seeing his face were convinced that he was none other than their king and indicate signs that they would be coming at night to take him out of the fort. Bor Sing was happy to see his own men who have managed to locate him and who were trying to free him. With night fall, the group then let loose a kind of a sitting frame that was made out of bamboo cane and U Bor Sing on seeing the same, sat on it immediately and was pulled away to safety and disappeared altogether from the place in the silent of the night.

This narrative does not give any details as to what happened after this incident at the fort and what steps did the Jaintia king took thereafter. Interestingly the historical chronological events that had taken place during the time of Borkuhain II were silent about these particular developments that had taken place during his reign. Perhaps this could be one instance to show case the importance of oral traditions and how the same can be reconstructed in order to understand as to what had really happened in the past.

The story of the Jaintia king and his relationship with his counterparts beyond his natural frontier would require further investigations and perhaps this narrative could provide a lead for further research work in the days to come.

NOTES

1. Tngensi Rynjah, *Ka History Ka Ri Khasi-Jaintia* (1562-1826), Part II, Vol. 1, p. 46.
2. Homiwell Lyngdoh, *Ki Syiem Khasi Bad Synteng*, p.71.
3. J. Bacchiarello, *Ki Dienjat Jong Ki Longshuwa*, p. 29.
4. Homiwell Lyngdoh, op.cit., pp. 79-80.

23

The Legend of *Ka* Iaw Chibidi

The folklore that is prevailing with the people refers to a woman who came from beyond the Kupli River and entered Jaintia Hills along with her children. The legend suggests that she has the capacity of accommodating herself in an earthern jar or *lalu*, by which she and her children were later known thereby giving rise to the emergence of the *Lalu* clan in Jaintia hills. The family prospered during the time when a powerful chief of the Malaniang clan held sway in Jaintia Hills. On the death of the king a civil war broke out, and the *Lalu* family, together with many others, moved across the river Kupli. It was here that they lived and prospered for some generations until they were affected by a plague which hit that area. The whole family was wiped out except for one girl and was known as Ka Iaw Iaw. She became the sole owner of the family wealth and her riches attracted many suitors and it became a problem for her to live in that place. She managed to flee away from that place and reached Jowai and then went straight to the priest house or the *lyngdoh*. Though the lyngdoh received her and allowed her to stay, his wife was thinking otherwise. She did not want to keep the girl and insisted on the priest that the girl should be sold off to anyone who wants to purchase her. Interestingly the Jaintia tradition and culture does not draw any reference to this kind of trade, yet it is still difficult to ascertain as to whether there was any form of slavery in the hills. However, a reference to *ki broo* or slaves in the colloquial speech of the people is very much there.

The narrative goes on to offer that the Langdoh under pressure from his wife, tried to sell Ka Iaw Iaw as a slave, but no one would offer more than twenty cowries (*chibdi*) for her. The priest then decided to keep her and took her back home, much to the displeasure of his wife. Out of gratitude for this kind gesture on the part of the priest, Ka Iaw Iaw told him about herself and decided to bring all her wealth from beyond the Kupli River to the priest house. This turn of events made the Langdoh and his wife happy and Ka Iaw Iaw was later married to their son.

Things were going on smoothly with the newly married couple, when some adventurers from beyond the Kupli decided to come over and to kidnap the bride for her wealth. The priest was able to get the information before hand about the intention of these people and made arrangements for their escape. The two were taken to a place near Nongkrem, called Sohphohkynrum, in the Khasi Hills and settled there. In this place Ka Iaw Iaw was paying every man, who was engaged by her in establishing a market, twenty cowries, or *chibidi* per day for their labour. In this place she was also credited for having introduced the art of smelting iron, and was said to have made various iron implements which she would export to the plain areas. The narrative has it that she also kept a huge herd of pigs which she fed in a large trough hollowed out of a *diengdoh* tree. For some years Ka Iaw *Chibidi*, as she was known by then because of her payments of *chibidi*, and her children lived in peace in that area.

News of her riches, her knowledge of smelting iron and her business in iron implements with the neighbouring areas spread fast. The oral narrative informs that the Ahom king Swarga Raja, U Long Raja (probably the Raja of Jaintia) and the Assamese Barphukan joined hands together and attacked this place Sohphohkynrum. Their intention was to take away the riches of Ka Iaw *Chibidi*. This compelled Ka Iaw *Chibidi* to run away from this place to Lyndiang-umthli, near Lyngkyrdem. But in this new place they found it difficult to stay together and the family decided to split into four groups for their own survival. One division returned to Jowai and came to be known as the Lalu clan, another group went to Nongkhlaw and became the Diengdoh Kylla clan; another group went to Mawiong and came to be known as the Pariong clan and the fourth went to Rangjyrteh and Cherra, where they established the Diengdohbah clan, and became afterwards one of the chief *myntri* clan of the state.

While trying to understand the legend of Ka Iaw *Chibidi* one cannot miss the process of migration of her ancestors and that they originally hail from the plain areas of North Cachar. This movement from one place to another have helped them to come into contact with people and to master the art of smelting iron. In the course of their relocation to save themselves, this new technology were imparted to the people by engaging them to work for the money paid *i.e. chi-bdi*. Her presence in the Jaintia Hills had played an important role in introducing the skill of smelting Iron and also to make iron implements. Further, the family's struggle for survival has also led to the emergence of a number of clans both in Khasi and Jaintia Hills.

Source: P.R.T. Gurdon, The khasis, pp. 63-65.

24

The Legend of *Ka* Taben and the Importance of the Market

The story of *Ka Taben* goes back to a time when nothing much can be substantiated chronologically, but reflects on the way of life and beliefs of the people. The Jaintias economic activities depended much on their geographical location and conditions. The geo-cultural factors play an important role and people belonging to different regions would specialise their arts and crafts accordingly. Their principal occupation was agriculture, but pottery making and basketry would also constitute an important aspect as their subsidiary activities.

There had been loud claims that the Jaintias were having trade relations with the people of the plains through Bengal, especially through Murshidabad during the Mughal period.[1]References about them extending their trade to Calcutta and Bhutan[2] have also been reported. Much before the coming of the East India Company into the Jaintia Hills, the Jaintia royal court was regulating the trade in limestone and coal and this had been the constant source of revenue to the King's exchequer. During the Mughal period timber was supplied to the plains. The principal market was Muladul where regular trade was conducted between the Jaintias with the Ahoms, Cacharis and Bengali traders. Trade in wax and ivory was also carried out.

This kind of activities throws much light on the socio-political life of the people and the kind of importance that the people were giving to trade and markets. Furthermore, their calendar which is being based on agricultural cycle and the eight-day week which is again based on market speaks volumes about the collective activities of the people.[3] It is in this context where importance was given to their occupation through their market deities that the legend of *Ka Taben* and her daughter Ka Rasong came into existence.

The Story Unfolds in this Manner

In the ancient times a woman by the name of Ka Taben, accompanied by her children came from the end of heaven and after wandering for sometime came to the land of the Plains people, referred to as the country of U Truh, bordering the territories of the Khasis and Jaintias. The woman offered herself to U Dkhar or the plainsman, as a household goddess, but he rejected her. She then move towards Khasi hills and offered to help them in their cultivation but they also refused saying that they could very well managed their own cultivation without her help.[4] She then moved towards the land of the Jaintias and after travelling with her children for few more days landed in a place called Rymbai. She offered herself to the chief, to help him supervise his daily economic and trade related activities. The chief who was looking for ways to boost his economic performance was impressed by this offer and agreed to accommodate her along with her children.

Ka Taben took over the task immediately and assigned duties to her children as to what they should do in the house of the chief. It is here that we have a reference to one of the daughters Ka Rasong. Ka Taben allocated to her the task of looking after the young unmarried folk and more importantly to supervise their daily labour. In addition to this she was also asked to handle their trading operations at the market place. This she did effectively and because of her help and blessing she was elevated to an important position in the family locally known as *U Tre- Richot*.

Ka Taben was able to perform her duties much to the liking of the chief and brought a lot of riches and prosperity to him and his family members. Her activities helped her to find an important place in the market and were respected by the people. After her death she took the form of a deity where she was propitiated along with the other market deities. This kind of reverence that was given by the people in time became a part and parcel of the cultural evolution of the people and is reflected through their rites and rituals.

References can be drawn during the Behdienkhlam festival, where the last day of the event would be followed by the Jowai market day. The *khnong* or the sacred tree would be kept in the market place before being taken to Aitnar or a place where the final rituals would be carried out on the last day of the carnival. This kind of practice speaks much about the importance of trade and commerce and for that matter the market place in the life and culture of the people. Market deities are also worshipped and propitiated with sacrifices of goats in places like that of Nartiang, Sutnga and Changpung. In Sutnga among all the riuals one of the most important is the market ritual called *Ka Pam Iaw,* where the Daloi and the Lyngdoh of Sutnga would conduct the ceremony and goats would be sacrificed. In Changpung the goddess Moochai which is again a market deity is being worshipped with thirty sacrifices. To the people the markets not only facilitate trade and commerce but would also play an important role in the socio-political life of a community. Many important decisions that would relate to the social and political life of the people would be taken in the market place only. They are also places of interaction where people from outside would also come into contact with the community. It is because of the importance and the role-played by markets that we find it to be mentioned in the oral traditions and folk tales from time to time.

Today, this kind of reverence is still being given to the different market deities and ceremonies are being attributed to them for the well-being of the people, the community, and the state.

NOTES

1. P.R.G. Mathur, Khasi of Meghalaya – Study in Tribalism and religion, p. 43,also in Soumen Sen, Social and State Formation in Khasi and Jaintia Hills, p. 68.
2. E.A. Gait, History of Assam, p. 358.
3. Martin P. Nilsson, 'Calendar', Encyclopedia of the Social Sciences, Vol. 3, p. 141.
4. Soumen Scn, Social and State Formation in Khasi and Jaintia Hills, p. 72.

25
The Mars of the Jaintias

In the *Pnar* context, the term *Mar* is being used to refer to those people who have gained status and recognition owing to their exceptional physical strength and endurance. Their prowess does not remain confined to a particular village but instead their name and fame would spread throughout the Jaintia kingdom. Till today these selected few are being regarded by the people with great esteem and there were times when special attributes are being given to them, which are again due only to demigods as can be seen in the Greeks, the Romans and Hindu mythologies. Though the narratives about these people are still very strong with the Jaintias, yet it is difficult to ascertain as to what extent they are true, without the element of exaggeration. The details of their works and achievements at best can be described today, to a rational mind, as nothing more than fiction and are coloured with fanciful ideas. But just like any other folklore, it is for the reader to understand the meaning and content of the same and also to differentiate between the myth and the reality. The intentions and desires of the people to generate such stories, so as to fulfil the needs of that period call for further introspection.

Whatever might be the interpretation and the judgement, which is again highly subjective, the story of U Mar Phalangki, the most prominent of them all and the other mars of the Jaintias runs as follows:

U MAR PHALYNGKI: Version I

The original name of U Mar Phalangki is not known. Though his clans name is very prominent, his first name had receded into the background when his people started to refer to him as *U Mar* because of his exceptional strength and deeds. U Mar Phalangki was a trusted lieutenant of the Jaintia king,[1] the chronology of which cannot be ascertained as yet. U Mar Phalangki is being remembered today for a number of stories that revolves round him. The most common being his erection of a monolith at *Iaw Mulong* in Nartiang and is known till date as *U Moo Syiem*. The Moo Syiem is the tallest and stands at 26'5" by 6'9" by 2'3" above ground.[2] The other monoliths in this place were erected by U Mar Phalangki and U Luh Langskor Lamare along with the inhabitants of Nartiang village between 1500 and 1835 A.D. to commemorate glorious events of the Jaintia kingdom.[3]

Since a considerable part of time had elapsed between that period and today without any documentation or otherwise, the writing will have to depend heavily on the oral narratives. Today the following version is being regarded as the most accepted, though without a chance for cross-examination in reconstructing the past.

U Mar Phalangki was a well built and had a very strong physique, intelligent and was much ahead than any of his counterparts during his time. These qualities had helped him to have an extra edge over his friends be it in the display of strength or in matters of the mind. With time he commanded respect from both far and near and more so from the people of his own village, Nartiang.

U Mar Phalangki got married to a woman from Ralliang village which is quiet a distance from his own. Going by the customary practices of the Jaintias U Mar Phalangki use to work and eat in his parent's house and in the night he would go and stay with his wife but would return early morning to his parent's house before sunrise. This practice is still in vogue till today. On one occasion when he was getting ready to leave his wife house in the morning, it poured heavily and U Mar was looking for something to cover himself from the rain. He then asked his wife to provide him with something and she feeling tired to get up because of the inclement weather jokingly told him to take one of the dolmens from the market place and to use it as his shield from rain. He took her word seriously and went straight to the market place and picked up the huge flat stone and proceeded for home. On reaching Nartiang the weather cleared and U Mar took the huge stone and placed it in the village market. It was said that the stone which he carried was the one on which the people of Ralliang use to carry out their rituals and sacrifices and was commonly known as *U Moo Niam Iaw* Ralliang. News spread about this act of U Mar Phalangki and a huge crowd gathered to see the same. There was a great deal of excitement with the people of Nartiang and were happy with the turn of events, for the prosperity of Ralliang depended much on that religious stone or *U Moo Niam*.

Meanwhile the people of Ralliang were shocked when they came to know about the deeds of U Mar. True to the belief and practices of the people, the village market that once use to see so much of prosperity and excitement during the weekly market days started to diminish with every passing week. This made the people sad and a meeting was called to discuss on the possibility of bringing back their *Moo-Niam*. The people then decided that it was important for them to get back their religious stone and this should be done at night at the time when everyone would be asleep and more importantly when U Mar would be in his wife's place. The plan had to be chalked out properly, as the people of Ralliang did not want any kind of confrontation with the people of Nartiang and more so with U Mar Phalangki. Accordingly the people of Ralliang entered Nartiang in the night and carried away the huge stone which rightly belonged to them.

The next day when the dawn broke, no one in Nartiang realised about the chain of events that had taken place in the previous night and the people were carrying out their activities as usual. It was only during the day that the inhabitants of Nartiang realised that the stone was missing and the rumour has it that the people of Ralliang have taken the Moo- Niam. U Mar was informed about these developments and he went straight to Ralliang during the day and in front of everyone took the stone without facing any kind of resistance from the people of the said village. The people of Ralliang were mere spectators about the turn of events. It was said that after this incident the people of Ralliang never even tried to take back the stone and it remained ever since in Nartiang. According to tradition the *Moo-Niam* brought a lot of prosperity to the people, whereas in the case of Ralliang, the market met its sad demise in time. Reference about the people of Ralliang lamenting about the loss of their Moo-Niam can be heard at times.

Another story that is still being remembered by the people about U Mar Phalangki was the erection of the *Moo-Syiem* or the King's stone at Nartiang. The stone was meant basically to commemorate the rule of the Jaintia kings and measures around 26'5" by 6'9" in breadth and 2'3" in thickness above ground. It is still difficult to ascertain as to whether the act of doing the same was done by U Mar Phalangki alone or whether he was assisted by U Luh Langskor Lamare, who was also regarded to be another *Mar* and close to U Mar Phalangki. Legends have it that U Luh Langskor Lamare had assisted U Mar Phalangki in a number of occasions in erecting the stones in the market place.

The traditional narrative offers that a large number of people had gathered around to witness the act of erecting the monolith. It was said that many attempts were made by him in erecting the huge stone, but every time it was done, the stone would never stand up straight. It would always tilt in an inclination and would never remain secure. U Mar Phalangki then decided to erect the stone on human blood and to this effect he dropped his lime container into the pit that was dug to erect the stone. He then called a small boy from the crowd to go down and to pick up the lime container. When the little boy bent down to pick the same, U Mar Phalangki carried the huge monolith and ram it down on the innocent boy. Blood splashed all around as the huge stone crushed the fragile body of the little boy. As for the crowd that had gathered there, they were taken aback by the act. The uneasiness of the calm was soon broken when the people realised that the stone had stood erect on human blood. Very soon there was a lot of merry making as the deed was completed. What was the need or the desire to fulfil such acts is not very clear. Does it have to do with the practice of human sacrifices of this place? or did it have to do with any special commemoration process? or with any specific religious act? These are some of the questions that have to be looked into in far greater details. Going by the nomenclature and the manner in which the stone is being regarded today it would clearly mean that it was meant for the King, as an act of glorification to the king and his *Raj*. Another area that needs to be looked into is the worshipping of the gods and spirits with human blood, in other words human sacrifice, which was common with the people of that place. This again would suggest that the sacrifices of humans as an act of divination was there with the people and stray references of the same can still be heard even today especially when we would relate to the human sacrifices that were carried out by the Jaintia ruler in the not so distant past. This became more pronounce when the Jaintia rulers were exposed to Hinduism and tantric cult from the plains.

Version II: U Bailon Khyriem

Apart from the above the Jaintia narrative do give a few more references about people with exceptional strength. It was said that before the British occupation of Jaintia Hills, there lived a *Mar* in the small village of Nangjngi who was tall and big build, to the extent that he would jump over small pine trees and would treat them like shrubs.[4] Specimens of some clothes worn by these Mars were to be seen in some villages of Jaintia Hills not very long ago dating to the 80's of the last century. It is difficult to say whether these clothes are still being kept by the people for posterity. The eyewitness accounts gives an interpretation that, three average Jaintia persons of present day would fit in those clothes.

The accounts about Bailon Khyriem, a *mar* from Um Latkhur village provides interesting information. It was said that on one occasion while he was going home, it started to rain. He took one huge slab of stone and used it as his rain guard and carried it all along the way. When the rain stopped he placed the stone by the side of the road between Thangbuli and Loom Chyrmit[5] which can be seen till today. A reference about his walking stick of four feet in length with a diameter of about 5" at the base is still to be seen in one of the households at Um Latkhur.[6]

References about the Jaintias building bridges across rivers with huge slab of stone or *Moo-Yieng-Kein?* are plenty. To this there is one old narrative which says that, once in Jowai there used to live two women with two sons each and that both of them were widows. The first one named her sons U Mak and U Maklai and the second one named them as U Luh and U Tapum. It was said that right from the time when they were small, these four have showed exceptional intelligence and strength. When they were at the prime of their age they decided to go and meet the king at Jaintiapur and to work for him. The king, who at that time was contemplating of improving the roads and building stone bridges over the rivers, was happy to make them work for him. Tradition has it that they were the ones who sculpted the sun, the moon and the king's pond with the elephant head and trunk at Rupasor. They have also erected a number of stones on the side which were meant as sitting place for the king and his ministers. After the completion of the bathing *ghat* they started bridging a number of river and streams, and one among the many was the stone bridge at Thloo-Mu-Wi. Local narrative offers that the four wanted to put a single slab of stone across the river, but while carrying the same the huge slab broke on the way and the place is known till today as *I Kyndon Mokhain*. It was after this that each one of them decided to carve a slab each, and to carry and put the same across the river. The first one was said to be of U Mak, since he was the eldest, the second that of U Luh, the third that of U Maklai and the fourth of U Tapum.[7] During the days of the Jaintia Raj the *Mars* used to complete the job assigned to them by the king in a short span of time. One can also refer to the stone bridges at Um Yaknieh and Syndai, the stone pools at Pdengchakap and not forgetting the host of other stones that were erected by the people at different stages. In short they can be referred to as *Ki Moo- Pynyein?*, *Moo-Knor*, *Moo-Boohnam*, or memorial stones; *Ki Moo-Kroh*, *Moo-Tpep/Moo-Niam* or the religious stones.

The Jaintias would also erect stone not only as a mark of remembrance to commemorate an act or an event but also to spite and also serve as a warning to many. This is being referred to as *U Moo-Pynboodnam*. A dolmen at Amlarem in the southern part of Jaintia hills would testify. On this stone the private parts of a lady with ill repute was carved out,[8] serving as a warning to the others who are involved in such acts. The stone carving was defaced not very long ago. The memorial stone or *Moo-Boohnam*, apart from the many which are basically associated with the political authority, the Jaintias were also known to have the same for a different purpose altogether, like in the case of *Ki Moo-Le-Phau* and *Ki Moo-Le-Spah*. In the case of the former which would literally mean the Thirty Monoliths, they were erected in the memory of the thirty husbands of a woman from Sutnga.[9] In the case of the latter, which would again mean as The Three Hundred Monoliths, they were raised in the memory of the husbands of a woman from Chken Pyrsit. The traditional narrative has it that, a few hundred years ago there lived a woman with three hundred husbands in this place and after her death, she was buried on this mound and each of her husband's placed a stone as a mark of respect to the deceased.[10] Prof. B. Pakem was of the opinion that the place was so called because an Amwi lady was once criminally assaulted by three hundred enemy soldiers,[11] and the stones were erected as a mark of remembrance to this lady who met such a fateful end. To this, there is another local story which narrates about the events that had taken place during the British period. This local narration mentions that, during the Jaintia Resistance of 1860-1863, around 300 British soldiers were killed in this place while they were trying to suppress the uprising of the people. As a mark of respect to the departed souls a stone each was erected on their behalf and, therefore, the term *Moo-Le-Spah* or the three hundred stones. Though such an incident finds no mention in the British records, yet the Memoirs of James Howard Thornton, who had taken part in the actual fighting, writes that this place saw considerable troop concentration.[12]

The Mars of the Jaintias were also known to have erected stones for resting purposes which are known as *Moo-Chong-Ngiah* like the ones in Khlieh Langcha, a locality in Jowai town. They were also credited to have erected a large number of stones in the market place which are known as *Ki Moo-Iaw*, like the ones in Nartiang and Nongbah. Apart from these, they were also credited for having huge slab of stones which were used as a wrestling ring, during market days like the one at *Iaw-Khyllaw* at Sutnga and *Moo-Sylli* and *Moo-Kor* for observation, serving as observatory post in times of war and stone fortress respectively.

Great personalities like that of U Railong, U Ran Niangti, U Bailong Khyriem, U Ter Kyndiah and U Khur Ralliang is still being remembered by the people, along with their deeds that revolve round them. Though detail information about these people cannot be easily attained, yet the story about the trial of strength between U Ter Kyndiah and U Khur Ralliang had filtered down through the writings of Mr. Paty Endrison Kyndiah.[13]

Version III: U Ter Kyndiah

According to the information available U Ter Kyndiah was born in Jowai and grew up in the traditional Jaintia way of life. Right from the time he was a kid, he had made his mark in different fields, and was quiet ahead from his own contemporaries, especially in size and in strength. He kept his hair up to his shoulder and would wear a ponytail neatly as was the customary practice during that time. He grew to a height of six an half feet and was muscular. He would wear a dhoti and a black jacket commonly known as *ka Iangki*. With time his name and fame started to spread to near and distant places, much to the curiosity of the Jaintia king (the chronology that could not be ascertained and some would place his period in the middle of the 18th century) who then invited him to visit the royal capital. There is no information as to what had really chanced in the meeting between him and the Jaintia king. Some narration has it that the king was very impressed by his strength and valour and decided to entrust on him the task of commanding his army.[14] If one would go by this assumption that he belonged to the middle of the 18th century, it would indicate that it was during the time of Bor Kuhain II (1731-1770) or Chatra Sing (1770-1782). Interestingly, the English East India Company carried out an expedition against Jaintia in 1774, which is again being referred by some as the First Anglo-Jaintia War of 1774, and there is no reference whatsoever in any of the British records about the activities or exploits of such a character even though the security of the Raj was at stake. On the other hand, information about the military campaigns, troops deployed and the Jaintia war tactics and the role of the Jaintia king is available and details can be extracted from the archives. What is more perplexing is the fact that if U Ter Kyndiah was actually made the commander or for that matter one of the commander of the Jaintia army, his role in the war should have filtered down in one way or the other. This kind of queries does not in any way minimise the narratives which still runs strongly with the people. What is found wanting is perhaps a proper understanding of the turn of events and to look into the other details more explicitly and to draw a connection between the real and the unreal.

Narratives about him suggest that on one occasion, U Ter Kyndiah had problems with the members of the royal family at Jaintiapur and decided to leave that place and to go to Jowai. On reaching a place called Rupasor near Syndai, a place where the king's pond is located which is about 360 sq.ft and carved out of a single piece of rock, with an elephant head sculpted on it[15]; U Ter Kyndiah out of anger broke the elephant trunk with just one twist and threw it at a distance. This elephant head without a trunk can still be seen till today. To this, U Primrose Gatphoh has a different version in his book *Ki Khanatang Bad U Sier Lapalang*, which was first published in the year 1937. According to him, U Ter Kyndiah was a merchant from Jowai who used to frequent the market in the plains of Jaintiapur. On one occasion while he was on his way to Jowai he fell sick from fever at Rupasor after having his bath on the king's pond. When he recovered from his sickness he was so angry with what had happened to him and blamed the pond for his sickness and took revenged by stoning the elephant trunk till it broke.[16] This version is quite different from what is being stated earlier and does not reflect much. Of late steps were taken to renovate the same, but the story around the elephant trunk is still being told and narrated.

Stories about his strength were told drawing inference from a cold winter day. On one such occasion while he was basking in the sun, the bamboos that were there near his place were preventing the sunrays from falling directly on him. Out of anger he uprooted all the bamboos that were there, pulling them just like weeds and throwing them to a distant place.

U Ter Kyndiah is being remembered today mainly because of the duel which he had with U Khur Ralliang, a mar from the village of Ralliang. As it is still today, the Pnars of Jaintia hills would have their market days spread all over the week in different places. The people of Jowai are very business minded and they used to frequent these markets in different parts of their kingdom. On one such market day in Ralliang the people of Jowai who were there to carry out their trade happen to hear the loud challenges put forward by U Khur Ralliang. The people of that place were scared of him because of his physical strength and his ability to inflict harm. An open challenge was put by him against the Jowai people, who were there to carry out their trade. U Ter Kyndiah who happened to be there in the market place on that day was impressed by the courage and strength of U Khur, but wanted to befriend him. U Khur on the other hand refused to acknowledge the same and instead put forward his challenge. The people of Jowai rallied behind U Ter and requested him to save the pride and prestige of the people of Jowai by accepting the challenge and to fight U Khur at the market place. Meanwhile the elders of the village tried to talk some sense into U Khur, but he was adamant and flatly refused to accept their opinion and stressed that he wanted to put an end to the proud bearing of the people of Jowai. At this stage U Ter also accepted the challenge and the fight was due to take place on that day itself.

Going by the convention of these kinds of fights, the same should be fought with bare hands and no arms are allowed by both the parties. The fight was to be conducted by a religious priest and instruction being given before the fight. U Ter Kyndiah was said to have said these words:

"Ym em u wan o cha Raliang wow pynmih iaklumar nea wow wan iachoh bru. Neibhah itea, sniaw o wa hakhmat u bru u Blai, nga ym ktah kti chwa o iei I iei I ileh, lait sad u wa em ki wa ktah kti chwa ia nga".[17] On translation it would mean, "I have not come to Ralliang to create trouble nor to challenge and fight anyone. In that sense I feel that in front of man and God, I will not touch the other person first, except that somebody touches me first".

The priest also put down his opinion with these words: *"Wa iwon iwoni wa mih na I ni I pynyoo bor ha pyrdi ki ni ki arngut ki mar, ieites toh u chim watoh I mon iong u Blai. Ym toh u ia boohnood boohkpoh I, iasih ia-choon ha pyrdi ki bru iong ki ni ki artylli ki chnong ki waheh, itea ka Ilaka Jwai wa ka Ilaka Raliang. Toh u pynneh pynsah iei maia kam chi paiu hei chongsuk chongsain kam wa jooh man chilynter. Toh u man I, I ia pynioobor iongiung iong lok hi eh, wom iahoi u booh syin? tyrkhaw".*[18] Meaning, "Whatever translates from the show of strength between the two mars have to be accepted as the will of God and that no enmity should be shown by the people of the two villages, *i.e.* of Jowai and Ralliang; that the love and respect should remain as before and should not end up in hatred with one another".

The people from both the sides who were present agreed that the fight between these two has nothing to do with them and that even after the fight whatever might be the outcome of the same, they would still carry out their life as usual, and that there should not be any kind of enmity between them. With this the trial of strength between the two *mars* started.

The challenger started the proceedings by drawing a line with his thumb on a slab of stone nearby and as he continued doing the same, smoke started to come out, much to the amazement of the onlookers. He then turned his attention to u Ter Kyndiah and said that the same fate was waiting for him when he will do the same to his body and also added that on which part of his body that he would like to start. U Ter pointed to his stomach and U Khur tore into his flesh and took a portion of the soft tissue from the stomach with his bare hands. To the surprise of many, neither blood nor anything ooze out from the body of U Ter and the wound that was inflicted on him also healed instantly in front of all the people. It was then the turn of U Ter and he asked for U Khur's hand as a sign of friendship but in that vindictive index, U Ter caught him by surprised by grabbing his hand and flung him high in the air, throwing him on the other side of the market. By the time U Khur landed, his hand was dislocated from his lifeless body. The people of Ralliang dispersed the moment they saw that their man had fallen defeated. U Ter, then turned to his people to follow him and they left for Jowai.

This narrative throws some light on the way of life of the people, their weekly hats, trade and how in the market days such events are also being held.

NOTES

1. P.J. Bazely (ed), *Meghalaya, Places of Interest*, p. 81.
2. S.N. Lamare, *The Jaintias, Studies in Society and Change* p.132.
3. P.J. Bazely (ed) Op.cit.
4. B. Pakem, 'The Megalithic Culture Of The Jaintia People Of Meghalaya' in*Golden Year Of Sein Jaintia, Sah Kynmoo 1947-1997,* p.12.
5. *Ibid.*
6. *Ibid.*
7. U Primrose Gatphoh, *Ki Khanatang Bad U Sier Lapalang,* pp.34-37.
8. B. Pakem, Op.cit, p. 9.
9. *Ibid.*
10. Shobhan N. Lamare, 'A Trip To Syndai' in *The Sentinel, 25:03:1995,* p.1.
11. B. Pakem, Op.cit p.9.
12. Shobhan N. Lamare, Op.cit.
13. Paty Endrison Kyndiah, 'Ki Mar Ka Ri Jaintia' in *YAKUNE* Vol. 1 No. 1 December-March 1999, pp. 7- 9.
14. *Ibid.*
15. S.N. Lamare, *The Jaintias:Studies In Society And Change* p.53.
16. U Primrose Gatphoh, *Ki Khanatang Bad U Sier Lapalang,* p.36.
17. Paty Endrison Kyndiah, op.cit.
18. *Ibid.*

26

The Origin of Rice Beer among the Jaintias

The use of rice beer among the different tribal groups of India is indeed something which is very common. In the context of the tribal societies of Meghalaya and the Jaintias in particular, the use of rice beer can be seen during the time of religious festivals or during any other secular practices. This has been part and parcel of the life style and culture of the people.

If one would look into the historical usages of the same it would be very difficult to trace down as to when the people have started using rice beer and also to make it a part of their cultural and religious practices. The tradition of the people speaks much about the manner as to how the knowledge of brewing the intoxicant came to be known and remained attached with the people.

The oral tradition of the people refers to a widow who had a number of children and meeting the daily requirements of her offspring's was a very difficult task for her. She would work in the field from morning to night and would take her children along. The elder ones in the family would assist her in her chores while the younger one would play around the work place. In the evening she would take them back home and prepare a meal for them. This had been her daily practices. What was unusual with the family was that the children would not go to sleep unless they were breast fed by their mother and this would also refer to the elder siblings too. This had been a major cause of concern for the woman. With no other alternative and seeing the kind of restlessness of her children if they were to be deprived of the same, she would bear in silent the pain and agony and would continue breast-feeding them.

On one occasion while returning from the field, she came across a plant which was not familiar to her, but got attracted to it and decided to take it home. On reaching home she did not know what to do with it but did not want to let go of it. To this effect she decided to seek the help of the local priest. The priest on seeing the plant sought for divination and after carrying out the rituals which were locally known as *I Lai Blai*, pronounced that the plant was meant to lessen her burden and that she will have to use it wisely. The oracle also revealed that the problem of breast-feeding would be things of the past. It further suggested that the plant should be pounded nicely and then left to dry for a few days. The dried substance was then to be soaked in cooked rice and the mixture to be left for three nights. Water was to be poured on the mixture and after mixing it thoroughly, the distilled liquid can be given to the children and this will keep them away from asking for breast milk. The women after listening carefully decided to carry out the same as she wanted to get rid of the habits of her children especially the older ones in the family.

She went home and did exactly as she was instructed and after a few days time she was able to give the liquid to her children, before serving their supper, starting with the elder ones. The children after having the liquid and their supper started to feel sleepy and soon enough fell asleep without asking for their breast feed. For a moment the woman was stunned by the turn of events and was very happy as it worked exactly the way it was ordained. She decided not to give the drink to the youngest of the family and after the usual breast feed the child went to sleep.

Time passed, and her daily chores of going to the field, preparing food and drinks for the children continued and comparatively much happier than what she had to go through before. This attracted the attention of her neighbours who were now curious as to how this change came about and what had prompted the same. She was then visited by a few and her story was narrated to them and also about how to prepare the drink. Those who were there got the opportunity to taste the drink and they liked the same. Slowly with time the news and method of preparing the rice beer spread to the others and people, both male and female, started taking the drink specially before sleeping as the drink made them doze off easily.

The mixture from the plant after pounding and processing is known as *U Thiat*, the distilled product as *ka sadhiar* and the left-over mixture after distillation is known as *U Kho-iang* and this is a very common drink among the Jaintias of Meghalaya till today.

(This information is being provided by Ms. Betty Laloo, Research Scholar, Department of Culture and Creative Studies, NEHU, Shillong).

27

The River Kupli

The river Kupli originates from the hills which lies to the east of Jaintia Hills, which forms a part of the boundary line between the Khasi and Jaintia Hills District and North Cachar[1]. It is the boundary between the area of the Pnars and that of the Hadems. Gurdon is of the opinion that the word Hadem is possibly a corruption of 'Hidimba', the old name for North Cachar[2].

From the time the Jaintia kingdom had extended to the plain areas, ideological influences were profound on the people especially on the ruling authorities. One of such influences was in the field of religion where the tantric cult had played an important role. This can be seen clearly with the worship and the kind of sacrifices that is being made to the river goddess. A traditional belief is still strong with the people and asserts that the goddess Kupli and her son U Iale[3], needs to be propitiated with human blood once a year. This had a lot to do with the practice of Human sacrifices among the Jaintias which was also carried out at the shrine of Jainteswari, at Nijpat, in Jaintiapur, and also at Nartiang. Right from the time when this particular ideology was accepted by the ruling family down to the time of Ram Sing II (1790-1832) there had been offerings of two persons during the months of November and December which coincides with the time of sacrifice that was carried out at Jaintiapur[4].

P.R.T. Gurdon who tried to look into this aspects of the sacrifices remarks, "There seems to be an idea generally prevalent that the Raja of Jaintia, owing to his conversion to Hinduism, and especially owing to his having become a devotee of the goddess Kali, took to sacrificing human victims; but I find that human victims were formerly sacrificed by the Jaintias to the Kopili River, which the Jaintias worshipped as a goddess"[5]. He further adds, "a special clan in the Raliang doloiship used to carry out the executions. It seems probable that the practice of sacrificing human victims in Jaintia was of long standing, and was originally unconnected with Hinduism, although when the Royal family became converts to Hinduism, the goddess Kali may easily have taken the place of the Kopili River goddess"[6]

The general practice during those days would revolve round the ceremony that was performed and carried out by the Brahmin priests at Jaintiapur after which the victims would be taken to the hills. They would be first led to Changpung market, where they would be given food of their choices and then from there to Sumer and finally to Iooksi, a village close to the river Kupli. Some distance from this village is a place where the stone meant for this kind of sacrifice is located. It is a flat table stone and the execution was carried out there. The sacrifice was carried out by the Lyngdoh, Sumer, Taket, Iape Pyrngap, Sariang, Pastein, and the papang clans[7].

After the execution the dead body and the heads were thrown into the river. The tradition has it that the rites and ritual meant for this kind of sacrifices was under the direct order of the king. The Brahmin priest would observe the immolation in the presence of the representatives of the king. There is also a reference to a particular clan "Mula Khara" who would volunteer to provide people meant for this kind of sacrifices[8]. P.R.T. Gurdon refers to them as Mugha Khara[9]. It was said that this clan passed out to extinction because of this act of theirs. When no volunteers were available the people responsible for this kind of acts would resort to kidnapping. There is also a reference where the king's slaves were sacrificed[10] and this would be carried out under the supervision of the royal family.

The water god is mainly referred to Iale, the son of Kupli and is still being considered as a very powerful god by the believers. The people believes that "he wears a crown on his head which looks like a turban with loops fore and aft more lengthy to the front side. His glittering ring is as large as a star. In his hand he holds a diadem. His height comes to over nine feet. He is never talkative but speaks very little"[11].

This narrative is being referred from time to time and one cannot say with authority as to whether these acts were committed or not secretly. Rumours would pour at different times and at different intervals about such acts but there is no concrete proof of the same. The common practice is to offer a goat for sacrifice and the same is being carried out by the people belonging to one of the clans mentioned above. It may be pointed out that one of the main reasons for the annexation of Jaintia in 1835, as claimed by the British was the practice and kidnapping of British subjects for sacrifice.

The ardent believer would still acknowledge the power of the goddess and that of her son and would prefer to offer his/her sacrifices to the goddess Kupli in the forms of rice, fowls or goats. The offerings would depend on the type of transgression that was committed and it had been customary for the traveller to confess their sins before they cross the river.

At present many of the Pnars would still regard the Kupli with reverence. They would not cross the river without performing a ritual and a sacrifice. The general practice would be to leave behind some rice or any other food items that the traveller might be carrying along. The people who use to ferry goods across the river finds it very difficult to do the same especially on the side of the Jaintias, when the coolies would refuse to go beyond a point and nothing will induce them to do so. The story is not the same though, on the part of the people of North Cachar, presently Karbi Anglong District. The Kupli is propitiated by rituals and sacrifices in different parts of Jaintia hills like that at Sutnga, Changpung, and Nartiang. At Nartiang a tank where sacrifices are regularly performed is called Ka Umkoi Kupli or Umkoi Bir Ympa, a pool symbolising the water goddess

NOTES

1. P.R.T. Gurdon, The Khasis, pp. 178-180.
2. *Ibid.*
3. H. Bareh, The History and Culture of the Khasi People, p.323.
4. P.R.T. Gurdon, op.cit., p.179.
5. *Ibid.* p.103.
6. *Ibid.*
7. H. Bareh, op.cit., p. 323.
8. *Ibid.*
9. P.R.T. Gurdon, op.cit., p. 179.
10. *Ibid.*
11. H. Bareh, op.cit., p. 323.

28

The Story of Midgets

The narratives about midgets or small people are very common all over the world. They are popular because of their jovial and friendly nature that they have to offer. Interestingly, in the case of the Jaintias, the midgets are being feared and do not reflect much on their jolly attitude and are being regarded as not amiable.

This is a story of a place close to present day Mookyndur, on the side of the National Highway 44 which is around 16 kilometers from Jowai. It is a small village popularly known to the local people as *Dong Rachi* and has its own story to tell. The people residing in this village and around it were known to be living with the Dwarf spirits who were locally known as *Lakweh*[1] (the meaning of the same could not be obtained). With time the place came to be known after the spirits and the old name was no longer used but not forgotten.

The people residing in this place were often tormented by the dwarf's spirits who were notorious to have some magical powers. These dwarfs were often playful but there were occasions when they would cause sickness on the people of that area. The oral tradition of the people offers that whenever the people happen to cross the pathways of the dwarfs they would be inflicted with sickness and in most cases the unfortunate people would become weak and with time become thin. There were occasions when the people would be in a state of delirium and eventually leading to mental illness. This had created a lot of hardship and the people were finding it difficult to avoid the dwarfs as sometimes they would just appears in places where the people were already there or in places where people are going about their daily chores. The people then started to think of ways as to how they would free themselves from the wrath of these spirits. In one of the gathering of elders, the thought of propitiating the spirits was discussed and was decided that a goat would be sacrifice as an offering to the spirits in order to cure those who were suffering from the wrath of the dwarfs. Accordingly the sacrifices were made and it continued for some time but the problems of the people did not seem to ease off. The perpetual fear of being hunted by these spirits created a fear psychosis on the mind of the people.

Things were turning from bad to worse when the numbers of mentally sick people were increasing. The only alternative left for the people was to leave the village and to settle in another place so as to get rid of the spirits of the dwarfs who were tormenting them with each passing day. To this effect the people relocated themselves to a place called Tyrchang, not very far from the old village. In this new place the people were said to be free from the clutches of the spirits and were leading a far healthier life than what they were leading before.

Tradition has it that the dwarf spirits would appear on a full moon night and would try to imitate the life style of the people. In this sense during the Behdienkhlam festival and especially on the last day when there will be a lot of merrymaking, these spirits would try to emulate the ways of the living, by engaging themselves in drinking and dancing. There are loud claims from the people of that area that some of them had the opportunity of not only hearing them sing and dance but indulging in drinking too. The general description was that these spirits would come out in groups and they would be walking and dancing by clapping their hands all along the highway and would pass by Ummulong village, which happens to be on their pathway.

This reference to Ummulong as a place where the dwarfs' spirits would frequent has faded out on the alleged reason that the number of vehicular traffic plying on the highway had increased many folds, much to the discomfort of the dwarf spirits and as a result has migrated to another area. The spoken tradition of the people would refer to a place near Nartiang where the dwarf spirit or the *Lakweh*, have migrated.

NOTE

1. Information provided by Mrs. Iris Lamare of Ummulong Village on 08:08:2010, 2 PM.

29
The Story of *U Thlen Lawania* and *Ka Blai Pynsum Kule*

VERSION I

This is one story which has been in circulation with the people of Ummulong, a place which is only 10 kilometers from Jowai the headquarter of West Jaintia Hill District. Though the place Ummulong is not far from Jowai or for that matter from Shillong, the state capital, we find that many of the stories relating to this place have not been much in circulation. One such story is about the *Thlen* or the serpent and *Ka Blai Pynsum Kule* or the water goddess. What is more surprising is the lack of interest on the part of the people, especially the youths to look into these stories, though the relevance of the same is still very strong, and the latest can be traced to the month of June 2010, where the people of this area and the followers of the indigenous religion *Ka Niam Tre*, carried out their rites and rituals so as to appease the super naturals and prayed for rain.

The narrative provides that in this village there is a small hillock called Lawania where a serpent god[1], locally known as *U Thlen* resides. What is interesting to note here is the belief with the Jaintias, which is still prevailing till today, is that the concept of *U Thlen* is mainly attached with the Khasis and not indigenous to them, though the story of a snake seeking human blood has never remain foreign to the Jaintias. Gruesome details of kidnappings and killings in the past in order to appease the snake would circulate in the hills and this must have permeated into the Jaintia society too, as the topography of Ummulong is close to Madur-Maskut and Khyriem. This description can be understood as a borrowed idea from the

Khasi hills and as stated earlier the story of *U Thlen* was never a part of the life and culture of the Jaintias.

According to this storyline *U Thlen* do not have any specific appearance but can transform itself into different shapes, though the general attribution towards it would be in a form of a snake. During a full moon night, the local people of the village claims that they would see a fire like burning and moving from place to place and this was supposedly to be the handiwork of the *Thlen*. The people regarded it as a spiteful spirit which can cause sickness and even death. It was held that nothing would happen to people who are strong spiritually and mentally, or *Eh Rngu*, as it is locally known. This concept of *Ka Rngu*[2] though still debatable has generated the element of fear to such an extent, that it had necessitated the appeasement of the snake.

Not very far from the hillock is a small river locally known as *Pynsum Kule* or a place where horses were usually given their baths. Tradition has it that, this river had got its name after the *Syiem* of Mylliem[3] and his men had bathed their horses in that place. It is quite interesting as to how the ruler of Mylliem would enter the Jaintia territory along with his men and bathe their horses in that particular place. Would it also indicate that there was an opportunity for the ruler of Mylliem to visit the Jaintia capital and on their way back stopped at this place to take rest? Alternatively, can there be some other explanation? Another version refers to the Raja of Madur-Maskut and his men bringing over their horses to the river and giving them their baths[4]. The second version seems to be more tenable considering the proximity of the territory of the Malangiang state to the Jaintia territory or for that matter even by the Khyriem chiefs. One fails to understand as to how and why the chief of Mylliem, whose territory was far apart should come all the way to Ummulong village and bath their horses there. That the practice of giving horses their baths in this place is confirmed by the name given, where *Pynsum* would mean bath and *Kule* would mean horse. However, the nomenclature of the place would require futher research and investigation.

The narrative offers that the river was inhabited by a benevolent goddess who would bless the people of that area. The river serves as a source of drinking water and also to irrigate the vast agricultural field of the people. Prayers and sacrifices were also offered to the river during natural calamities or to cure someone from sickness. The sacredness of the river was maintained and people from distant places would come to offer their sacrifices and in return expected to be blessed.

The stories of the two supernatural entities took an interesting turn with the narrative suggesting that *U Thlen* fell in love with the river goddess, *Ka Blai Pynsum Kule*, and use to visit her from time to time. Even though the two places were far apart, with a distance of not less than two kilometers, yet the *Thlen* would express his love by frequenting the river goddess. In time she developed a liking for him and got married. Out of this union, children were born to them and they stayed with their mother *Ka Blai*. The tradition is silent about the role of the offspring but adds that *U Thlen* preferred to reside in his own abode and would visit his family from time to time.

Time passed by and soon enough the goddess or *Ka Blai* came under the influence of *U Thlen* and started developing his traits. She was no longer kind to the people of the village and this can be seen from the number of sickness and death that was taking place in the hamlet. Some of the villagers would even get drowned in the river, something which had never happened before, or would face an unknown and unexpected sickness and would eventually die. These developments compelled the people to keep themselves away from the river and would restrain their children from venturing near it. Till 1970's and early 80's the elders of the village would forbade their children to go out for a swim in the river.

These developments had brought a lot of uneasiness and fear in the village. The people started to think of a way as to how the two deities could be appeased. After a lot of deliberation the people came to a conclusion that they should offer sacrifices to both of them and accordingly the selection of animals was also to be carried out. A rooster, a pig, and a pair of yellow hens were selected for the said purpose based on the events and requirements. One cannot miss the influences of other traditions in this narrative. The use of blood to appease, and the selection of animals, in so many ways talks about the permeation of thoughts at the ideational level and how the same is being accepted and applied and then spread to the general public.

After the offering of sacrifices, normalcy returned to the village once again and the position of the goddess or *Ka Blai* was now elevated as the protector of the whole village, but not forgetting her role as the provider too. As for the case of *U Thlen*, being satisfied with the offerings, he had sober off to a great extent and do not disturb the daily activities of the people of the village. The kind of importance and relevance that the people etill have to these deities was seen on the month of June 2010, when sacrifices were offered to *Ka Blai Pynsum Kule* due to the absence of rain which had brought hardship to the people of the village. A day after the sacrifice, the village witnessed a downpour. This in so many ways had reinforced the belief and ritualistic performances of the people of the village.

A sacrifice of once in ten years was performed for both *Ka Blai Pynsum Kule* and *U Thlen Lawania* at *Pynsum Kule*. Today the river is treated with great respect and some form of sanctity is associated with it. Out of fear the people would refrain to do anything that would desecrate it and the legend lives on.

VERSION II

This particular narrative goes back to the time when the kingdom of Madur-Maskut (whose chronological time frame can be traced back to the 15th and 16th centuries) was having its considerable influences in the hills of this region. The story offers that on one occasion, the ruler of this kingdom (whose identity is not known) passed by this area and on reaching the stream his horse could not go further because of severe pain on his fore leg. The king got down from his horse and tried to figure out the root cause of the problem but could not figure out. He thought that the best way was to take the animal near the water so as to allow the horse to drink some and also to rest for some time[5].

It was during this time when the king was pondering as to what to do next, that an idea came to him, that perhaps it would be wise to massage the leg with cold water from the pond so as to bring some relief to the animal. With this thought he took the animal slowly to the stream and started massaging the leg. What happened next was much to the disbelief of the king. The horse was able to stand up without any difficulty and manage to come out of the pond by itself. The king was taken aback by these strange developments and thought that the water had some magical properties. He continued with his journey to his capital Madur and shared the news with his people, much to their amazement.

A few days passed and the horse started to show signs that it was having problems. The king then thought that perhaps it would be wise to take the horse once again to the pond and to give him a full bath and also to see that whether it would work this time as it had happened in the first instance. On reaching the place and while taking the animal to the stream, out of a sudden the goddess emerged from the water and spoke to the king telling him not to venture into the water. She advised him to get first a pair of yellow hens and also to call the people of the village to witness the sacrifice of the same in her name. The king at first was stunned by the appearance and command but complied with the same. A good number of villagers assembled and it was here that a kind of a pledge was made between the goddess, the king and the villagers. The goddess assured that she would be their protector from pain, suffering, and diseases and in times of drought, she would bring relief, provided they offer their sacrifices to her. The idea of providing water to the area acquires its significance from the fact that the entire tract is a fertile valley and the *Pnars* of that area had being carrying out their wet rice cultivation right from the olden days. She also asked that once in ten years a goat, a pig and four cocks have to be offered to her, failing which she will not protect them from their sufferings. She then asked the king of Madur to spread five beetle leaves and a beetle nut and then to tie a knot on a blade of grass to serve as the symbol of understanding between her and the people at large. The king then placed the hens on the local leaf called *ka sla lamet* and a particular tree *U snin* was placed on the side and the sacrifice was offered to the goddess.

Since nothing was known about the goddess before this incident, the people would refer to her as *Ka Blai Pynsum Kule* or the goddess of bathing horses, drawing references from the incident with the ruler of Madur and his horse. Right from that time till today, the name remained and the people would still refer to the goddess and this place with a lot of reverence. The people of the surrounding areas who are the ardent followers of the *Niam Tre* still believes in the healing power of *ka Blai*

Pynsum Kule and would be seen frequenting the place to have their baths in case of sickness or otherwise[6].

The second version in a way opens up new queries and gives a fresh insight about the events and things in the days of the past. The reference to the king of Madur-Maskut moving in the areas of the present village of Ummulong would invite a considerable amount of introspection as to the fact that how an independent ruler like the king of Madur would move into the territories of the Jaintia king and that too without any followers. Ummulong as a place is a fertile region and as mentioned earlier, cultivation has been carried out extensively in the olden days which can be seen even at the present day. The fertility of the soil must have attracted the attention of the rulers of that time more so when this village is at the border regions of both Jaintia and that of Madur-Maskut. The oral tradition of the people does not throw any light whether that area was under the influence of the Malangniang rulers and the reference to the ruler of Madur-Maskut in the narrative of the Jaintias has added another dimension to the political history of that time.

In addition to the above, the other questions that would come to the mind is that, in what way would it be possible for the king of Madur to sway the people of the village to come and assemble and to be part of the sacrifice if he had not had already a considerable influence on them? Why do the people of this place still refer to the king of Madur- Maskut, though of course in the first version the king of Mylliem was refered to? Strangely enough, the oral tradition of the people does not have any reference to the Jaintia king. Can this serve as an indication that the area that is being referred to here was not part of the Jaintia kingdom at one particular period of time? But if one would try to look into the composition of the village and the settlements of the clans, there is no doubt whatsoever that it was a *Pnar* village and the founder clans can be still be identified. There is still a strong reference to a hillock known as *U Lum Soo Yung*, which was supposedly said to be the place where the first settlement took place.

From the historical point of view, this particular oral narrative gains in its importance as it brings about a number of queries about the geographical extension of the Jaintia kingdom and that of Madur-Maskut, and how the same was lost and then brought under the control of the other power at different point of time in the 15th and 16th centuries.

NOTES

1. Betty Laloo, *Folktales of North-East India*, Don Bosco Centre for Indigenous Cultures, Sacred Heart Theological college, Shillong, pp. 106-109.
2. *Rngu* as understood is the invisible part of the self which remains throughout so long the person is alive. It has both is positive and negative energy. A person with a strong *Rngu* is said to be strong mentally and spiritually and no harm can come to him, but it is the opposite otherwise. Sometimes *Rngu* depends on the situation or the environment in which a person is placed in and the same person can be exposed to both the elements.

A person with a strong *Rngu* can be vulnerable also at times and for this a ritual and sacrifice has to be carried out to regain back the lost invisible self i.e. the *Rngu*.

3. Betty Laloo Op.cit, p. 107.

4. Interview with Edmond Lamare, Research Assistance, Department of History, NEHU, Shillong on 29 July 2010

5. This information is given by Mr. Edmond Lamare, Research Assistance, Department of History in his interview with Mr. Lieh Shylla on 08:08:2010, a member of the *Niam Tre* of Ummulong village and is also one of the leader in carrying out the ritual and sacrifices for the goddess in this narrative, the latest being on the month of June 2010.

6. Information gathered from Mrs. Iris Lamare, Ummulong, 07:08:2010.

30
The Sung Valley

The Jaintias were known to be expert in wet rice cultivation. The Sung valley is one among the best places for wet-rice cultivation in Jaintia Hills. It is located in the western part of Jaintia Hills bordering East Khasi Hills and off the National Highway 44 on the Shillong-Jowai Road. Some parts of the valley fall under East Khasi Hills. It is a fertile valley and people from nearby villages would cultivate rice along with other food crops and vegetables like that of maize, pulses, potatoes, tomatoes, cabbage, cauliflowers, beans, radish, chilly, ginger, brinjal, pumpkins, mustard leaves. Fruits such as oranges, mangoes, bananas, *amla*, grapes, *sohphlang* (tuber belonging to the ginseng family) and *sohptet* or wild apple are also found in abundance. It is interesting to know that this valley is popularly known among the Jaintias as the *Pliang-ja* or the rice- bowl of the Jaintias.

The oral tradition which circulates in this area speaks much about the nature of the soil and its fertility and would draw references to the *Dieng-iei*, which is found in the Khasi oral narratives. It was said that the tip of the tree fell on the valley and this made the valley to be very fertile. Another version would refer to the roots of the *Dieng-iei* which scattered to different directions and that some of the roots were accidentally dropped at the Sung valley and in the process making the land fertile. These versions of the oral tradition does not throw sufficient light on the manner in which the branches fell or for that matter how the roots spread and were dropped by whom, but interestingly would seek to explain the fertility of the soil by linking it to the *Dieng-iei*.

Another story which revolves round the Sung Valley is the reference to the *Lakweh* which was believed to be a spiteful spirit. This story does not refer to the fertility of the soil but nevertheless it takes about the valley. Tradition has it that the *Lakweh* would be seen around in a dwarf figure shape and would reside with her family in a particular locality of Mookyndur which is now known after her. True to her nature the *Lakweh* would never allow the people of that area to live in peace and would torture them in many ways and one of her favourite hunting ground would be the Sung Valley, where she would destroy the crops of the people especially when it was time to harvest. She was happy to see people suffer by her actions. Attempts were being made by the people of that area to trap and destroy her but to no avail.

When this continued for a long time and tired by the hardship and evil designs of the *Lakweh*, an old man from the village decided to take up the challenge to confront the latter and to put to an end the sufferings of his people. Before leaving for the valley, he took a mirror, tied it to a bamboo pole and headed towards the direction of the Sung Valley without explaining much to his people about his ideas. None in the village were sure as to how the old man would go about in capturing or driving away the malicious spirit. The old man was hoping of meeting the spirit and soon enough he got the opportunity as the Lakweh was already waiting to way laid her first victim on that day. On seeing her, the old man asked as to where she was going, to which she replied that she was on her way to trap and eat humans. The old man was not dissuaded by her threat and kept on walking. She then asked him as to where he was going. He replied that he too was on his way to trap evil spirits and that he had already got one with him. The *Lakweh* was taken aback by this answer and wanted to find out whether the old man was telling the truth or not. She demanded that he should show his spoils to her. The man then took out his bamboo pole and held it in front of her. What she saw drove a sense of fear as the face looked familiar to her. She thought that it was her daughter. She realised that she had underestimated the power of the old man and begged him to release her daughter. The old man was adamant saying that it was not an easy catch for him. What followed next was a series of requests and pleading to free the image from entrapment. The old man agreed on a condition that the *Lakweh* will have to leave the place and not to torment the people anymore or he would come after her and her family once again. After this incident the people of that area were able to go back to their fields without fear of being watched or tortured by the malevolent spirit.

Interestingly the people of Mookyndur have named a locality after the Lakweh and are residing there without fear and would consider the entire narrative as an thing of the past with no relevance in the present day context.

31
The Syndai Cave

The Syndai cave or Ka Krem Mahadev can be said to be one of the important caves in Jaintia Hills and that it had attracted the attention of many. The cave has been explored by many adventure lovers over the years. Apart from the natural surrounding and its location, the presence of a Shiva linga and a number of Iron Tridents meant for religious purpose is an added attraction. One would be curious enough at this point to ask as to why a Shiva linga and iron tridents would be scattered in the entrance of a cave, in the jungle, and that to in this part of the world.

Going by the oral tradition of the people, the cave was visited by Lord Shiva or Mahadev, as he is being locally referred to, in the remote past. This concept perhaps has a lot to do with the coming of Hinduism into the hills and the kind of influences that it had on the people of this area. Syndai for that matter is situated in the southern part of Jaintia hills and has close proximity to the plains of Jaintiapur, the then capital of the Jaintia king. Perhaps the kind of trade connections and the importance of this region, had in one way or the other brought along with it the Hindu ideas, beliefs, thoughts and practices, and in the process a sizable section of the local population was brought under its fold. Since it was the religion of the ruling family, it had been fashionable on the part of the local population to identify themselves with the religion of the ruling authority.

The oral tradition refers to the coming of the people from the plain to the hills and to worship inside the cave. A Brahmin priest use to reside in the cave during the winter months for the purpose of carrying out rituals for those who would come to worship. It may be pointed out that very close to the village is a carving on the rock representing a frog, whose feet are tied with chains. The people believe that the frog represents the earthquake and it was in this place that the earthquake was tied in order to prevent it from doing any more damage to the people and place of this region. If a reference to this earthquake would go back to the events of 1897, then the carving on the rock can be said to be of recent origin on historical terms. The people still believes that if it had not been for the chains, then the earthquake would have brought about more damage to this region.

32

The Tiger-Shaped Monolith

This story revolves around Sutnga village and talks about a minstrel who had a chance encounter with a tiger. The minstrel was in a habit to move from place to place and would play his flute much to the amusement of the villagers. On one occasion as he was passing by a dense forest to go to another village he came across a tiger that was ready to pounce on him. The man pleaded with the tiger to spare his life but the tiger did not listen to his request.

On this the minstrel begged the tiger not to kill him immediately and suggested at the same time that each should try to prove of his superiority over the other in his own way. The minstrel then proposed that, he would try to make the tiger dance to the tune of his flute and if he should fail, then the tiger can do whatever he wishes with him. The tiger agreed to the challenge and the minstrel started playing his flute and very soon the tiger was standing up on his hind legs and started dancing. The tiger was confused as to what was happening to him and it appeared that he was not in control of his senses. His entire anatomy was responding only to the sound that was coming from the flute. This went on and the tiger was getting tired. He told the minstrel to stop as he was being cheated by him and that he was having no intention to dance. The man pretended as if he had not heard the tiger for he knew that he would be in danger once he stop playing.

In the meantime the tiger was losing strength and getting exhausted because of the continuous standing and jumping. His feet could not support his weight anymore. The tiger then requested the minstrel and pleaded to stop playing over and over again but like before he pretended as if he did not hear him. Out of exhaustion the tiger died while he was still on his feet and eventually turned to stone.

The oral narrative subscribes to this story about the presence of the tiger-shaped monolith in this place.

33
The Vision of Johumbon

A specific time frame cannot be ascertained to this man Johumbon, but going by his dreams and visions it would appear that he made his mark in the 17th century. Johumbon was born and brought up in Changpung, a small village in present day Jaintia Hills. This narrative about this man and his vision speaks much about the role of man towards his environment especially in his relationship towards the lesser creations. It is somewhat thought provoking, should the events turn out the way that it was supposed to be in the story. What is amazing is the fact that the people in the past have already understood the manner in which man is treating the surrounding and how nature was trying to convey a message through the language of the animals and to correct the wrongs that was meted to them. This story runs in the form of a metaphor.

Johumbon was no extra ordinary person in the village. Like any other person, he would attend to his daily chores and performs his duty. On one occasion he fell ill and it took a very bad turn when he lost consciousness and was in a state of hallucination. He was in that condition for five long days and the people around him were just waiting for him to breathe his last. After the fifth day he regained consciousness and his health started to show signs of improvement. He surprised the people around him when he narrated to them about his strange encounter in those five days of delirium.

Johumbom informed that he saw all the gods and goddeses had congregated themselves on the banks of the river Kupli and animals of different types were also present except for man. In this meeting the animals pleaded to the gods that man had not only brought problems on them by hunting them down for food, but many of his sins and bad deeds have to be borne by the animals in the form of sacrifices that he has to offer to the gods and goddesses. The animals who spoke the most in the congregation were the goat, the cow and from the bird family the hen and the rooster. The other present there also talked about the excesses that was carried out by man and resolved that for this misdeeds on the part of man, he should be banished forever from mother earth. This according to them was important so as to maintain parity and balance. If this was not done then the wanton destruction that was carried out by man would increase manifold in the days to come, so as to make the extension of life on earth impossible, save his kind. The river goddess Kupli along with the other deities gave a patient hearing to the complaints of the animals and found it very difficult not to agree to what the other had to say.

One of the gods present there, acknowledge the excesses committed by man but at the same time referred to the others who were trying to live a right kind of life. It was further asserted that for the crime that was committed by a few, the whole of humanity cannot be punished, for there is still a lot of good with many. The god then pleaded to the other gods and requested them for the preservation of the human race. The question on the total annihilation of the human race has proved to be a very difficult proposal to the gods and goddesses to accept. By this time the general opinion appears to be a divided one and both the gods and animals decided to call upon the supreme deity to give his verdict. They decided to call upon the supreme deity *U Blai Iale* or the god Yale, who was again the son of the river Kupli.

It was not an easy task for *Iale* and after much thought, he gave his own opinion and observation to the problem. He concluded that man cannot be annihilated but then his truths and vices should be weight properly. He suggested a way as to how the same could be judge and to be understood. He instructed that both truth and vices should be filled up in two cups. He then ordered the two cups to float and to rise to the heavens. The gods and the other animals present there were curiously watching the developments and to their astonishment saw that the cup with vices was rising higher. However, when the two came back to earth, they realised that the quantity of truth in the cup remained the same where as that of vices had reduced considerably. When the gods saw these they unanimously rejected the demands of the animals and concluded that the human race will have to be preserved.

This narrative gives an impression as to how man was saved from destruction with the intervention of god Iale and it was because of his wisdom that mankind is still around on this planet earth today.

34
U Muior Dkhar

One of the oldest villages in Jaintia Hills is Nartiang. The story of this place goes back to the time when the king of Sutnga decided to shift his capital to Nartiang. Apart from the many stories that this place has to offer, this time Nartiang has to share about a sad incident that had taken place during the days of the Jaintia Raj. Reference is being drawn to one U Muior Dkhar from Sutnga village. Sutnga for that matter was the ancestral home of the Jaintia kings and they were ruling over the hills from Sutnga only. It was only with the expansion of the Jaintia territory that the rulers found it necessary to shift their capital to Nartiang and from that time onwards it remained as the summer capital of the Jaintia kings.

U Muior Dkhar was said to be one of the deputies of the Jaintia king, the chronology of which remains baffling. However, the oral tradition does not provide any information about the period or the name of the ruler that he was serving, but asserts that he was attached to the Royal family. U Muior hails from Sutnga village and was regarded by many as a man of many qualities. He was known for his feat on carrying out operations on the sick, a man with super natural powers who could do magic and call upon the gods and also having the knack to counterfeit the *Kattra-taka* or the Jaintia currency. It was this last quality of his which landed him into trouble with the Jaintia king. The king had been nurturing a thought as to how to get him assassinated. This proved to be a difficult task for him because of the magical prowess that U Muior was having and any attempt to harm him might just boomerang on the person who was intending to do so. This made the king to be cautious but the thought of eliminating U Muior growing more powerful with each passing day. Here one fails to understand as to why the king was bent on to assassinate U Muior, a man with qualities to heal the sick and whether his knowledge to counterfeit the Jaintia currency would really amount to such deeds. Though the oral tradition is silent about the other aspects, perhaps there could also be some other reasons as to why the king was so determined to have him eliminated.

U Muior was also credited to have dug two lakes in Sutnga village. These lakes are square in size and are eighty feet long. The purpose for digging the same is not known, but going by the oral tradition of the people, many of the people in the past they would dig ponds and lakes in order to show their strength and prowess. It was said that his intention was just to dig one lake but then when the first lake was completed it was taken over by the village for community purposes. This hardened the stand of U Muior and he decided to dig another lake slightly ahead of the previous one. These lakes can be seen till these days, but because of the lack of maintenance be it on the part of the local people or the authorities they are in a pretty bad shape.

The narrative of the people would refer to an incident in Nartiang, when the king invited his subjects to take part in a competition that was especially arranged for them. This event was carried out in order to lure U Muior and then to trap and kill him finally. A huge pit was dug and anyone who would dare to go right down to the bottom of the pit and then to climb out would be given huge prize money. Many were tempted but no one dared. Finally from among the crowd U Muior appeared and accepted the challenge. U Muior had gone to Nartiang from Sutnga not really for the competition but to pray and call upon the gods as part of his usual practices, pertaining to his super natural strength. When he heard from the people of Nartiang about the competition he thought of giving it a try, much to the delight of the people and especially the king who was looking for a way to get rid of him. The king on the other hand had always anticipated that there would be no one except U Muior who would accept this challenge, and true enough when the latter accepted the challenge, it all happened just the way the king wanted it to be.

On the appointed day of the competition, people from different parts of the kingdom trickled to Nartiang to witness firsthand about the entire event. Many came to see the pit but the depth of the same had made many to keep them away from it. Then it was the turn of U Muior. He had a good look at the pit and after a while took his *tapmohkhlieh* or shawl and kept it on the side of the pit and without any sign of hesitation, and to the surprise of many, jumped into the pit. The moment he disappeared into the darkness of the pit, the king who had already kept his men waiting, ordered that the pit should be filled up with lose earth that was kept on the side. U Muior was buried alive because of the enmity which the king had towards him. The killing was done in a treacherous manner. The people who were around were dumb struck and did not dare to say a word. They remained as silent spectator to the evil deed that was committed.

It was said that right from that time onwards, the members of U Miour family, who were having the gift of divination, whenever they were asked to talk or call upon the gods, they would automatically clear first their eyes, as if something is preventing them from seeing. This according to them was a strong reminder to the members of the family that U Muior was buried alive and that when someone clear their eyes it would signify that they were trying to clean the loose earth that was thrown on U Muior so as to be able to see better.

Glossary

Bneiñ : Heaven

Behdeinkhlam : A festival of the Jaintias to drive away plague

Dakha : Fish

Dalloi : Chief

Dakha Khlur : Star fish

Hukum : order

Iaw Bei : Ancestral mother

Ka Tangnoob-tangjri : Huge tree

Ka Rym-Aw : Mother Earth

Ka Syiem : A Queen

Kongmai : Elder sister

Kur Soo-Kpoh Khad-ar-Warnai : Clan cluster

Khloo Blai : Forest of the gods

Khuri : cup

Ki Blai : gods

Kwai : Betel nut

Masi : Cow

Niaw Wasa : Seven huts

Niam Tre : Indigenous Religion of the Jaintias

Pnars : Jaintias of the upland region

Pyrthai : Earth

Puri Blai : fairies

Sein Khynroo-Khyllood : Youth organisation

Sein Raij : Indigenous organisation

Sla : Leaf

Sla Khyndaw : Earth surface

Soo-Iung : Four houses

Soo-Kpoh : Four wombs

Tymmen : Old

U Khadynru Wasa :Sixteen huts

U Tre Kirot : God

U Lakriah : Messenger of God

U Loom : A Hill

Wait : Dao/Machete